Notebook of Medical Physiology:

Also in this series

Notebook of Medical Physiology: **Cardiopulmonary**
Notebook of Medical Physiology: **Gastrointestinal**
Notebook of Medical Physiology: **Renal and body fluids**

To Ailsa

Notebook of Medical Physiology: Endocrinology

WITH ASPECTS OF MATERNAL, FETAL AND NEONATAL PHYSIOLOGY

A revision text for candidates preparing for examinations in basic medical sciences; including multiple choice questions

Ross Wilson Hawker

MD(Syd & Q'ld), PhD(Q'ld), FRACP, FRACS(Hon), FFARACS(Hon)

Professor of Physiology, Department of Physiology and Pharmacology, University of Queensland, Brisbane; Visiting Professor of Physiology, National University of Singapore, Singapore; Member of Board of Examiners, Faculty of Anaesthetists, Royal Australasian College of Surgeons; Formerly Member of Board of Examiners, Royal Australasian College of Surgeons; External Examiner in Physiology, National University of Singapore; External Examiner in Physiology, University of Malaya, Kuala Lumpur

SECOND EDITION

CHURCHILL LIVINGSTONE

EDINBURGH LONDON MELBOURNE AND NEW YORK 1984

CHURCHILL LIVINGSTONE
Medical Division of Longman Group Limited

Distributed in the United States of America by Churchill Livingstone Inc., 1560 Broadway, New York, N.Y. 10036, and by associated companies, branches and representatives throughout the world.

First edition 1978
Second edition 1984

ISBN 0 443 03188 6

British Library Cataloguing in Publication Data
Hawker, Ross Wilson
Notebook of medical physiology: endocrinology. — 2nd ed.
1. Endocrine glands
I. Title
612'.4 QP187

Library of Congress Cataloging in Publication Data
Hawker, Ross Wilson
Notebook of medical physiology
Includes bibliographies and index.
1. Endocrinology — Outlines, syllabi, etc.
2. Maternal-fetal exchange — Outlines, syllabi, etc.
3. Fetus — Physiology — Outlines, syllabi etc.
4. Infants (New-born) — Physiology — Outlines, syllabi, etc. 5. Endocrine glands — Diseases — Outlines, syllabi, etc.
I. Title.
QP187.H42 1984 612'.4 83-21052

Printed in Singapore by Selector Printing Co (Pte) Ltd

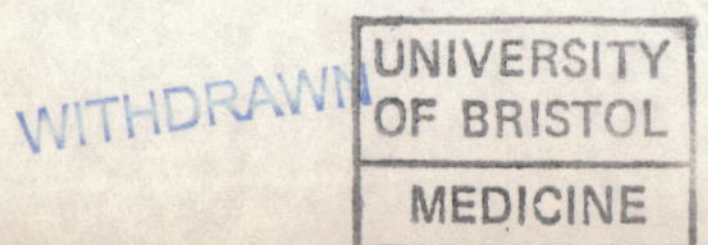

Preface

This is a revision text for students preparing for examinations in medical/surgical science. It is aimed primarily at postgraduates undertaking FRCS, FRACS, FFA, MRCP, FRACP and cognate professional examinations, and secondarily to medical students in both their pre-clinical and clinical years.

The style is unique and will not please everyone: it is deliberately cryptic and concise and attempts to compress large areas of physiology into readily assimilable facts and concepts with a minimum of description. This is illustrated by the many flow-diagrams and summaries. This didactic approach stems from over 30 years' experience on examination-orientated courses for the FRACS, FRCS, FFA and FRACP.

The self-assessment material in this book has several educational objectives. Firstly, many of the MCQ hopefully are of sufficient quality to merit their use by professional examining bodies, either directly or after further review. Secondly, a considerable number have been constructed to extend the material covered in the text or have some intrinsic teaching quality about them, which makes them unsuitable in their present form for examination purposes. Thirdly, some are intentionally provocative and are designed to prompt discussion by the reader with fellow students or tutors.

The format of the first edition has been retained and, in addition, concise overviews of some aspects of disordered function have been included in the chapters on the anterior pituitary, posterior pituitary, thyroid, adrenal cortex, pancreas and parathyroids. This new material has been included in an attempt to illuminate the clinical relevance of basic aspects of endocrine physiology and, thereby, to assist medical students further in their comprehension of clinical endocrinology.

Most of the chapters have been rewritten; many new teaching diagrams and multiple choice questions have been added; and the reading lists have been updated.

It is pleasing to note the book has been translated into German.

1984 R.W.H.

Acknowledgements

I am extremely grateful to Dr S.W. Manley for his constructive criticism and helpful suggestions in the compilation of the second edition of this book. His valuable assistance with the other three volumes in the series *Notebook of Medical Physiology* has previously been gratefully acknowledged.

I wish to thank Dr H.M. Lloyd who contributed the chapter on the parathyroid glands; and many colleagues who helped with the new material, especially the sections on clinical endocrinology.

Finally, I am also indebted to Mr D. Sheehy, Graphic Design Artist of the University of Queensland, for his excellent illustrations; to Dr B.G. Campbell, who prepared the index; and to the typists, Yvonne Shields and Jan Frousheger, for their skill and forbearance during the production of this manuscript.

Contents

1. Anterior pituitary

ANATOMY

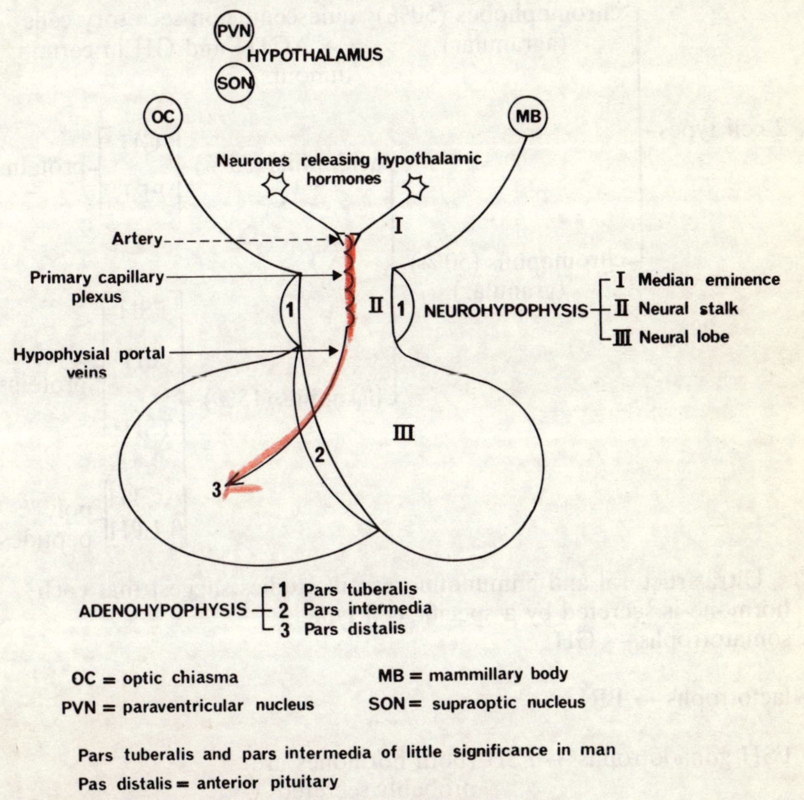

Fig. 1.1 Anatomy of pituitary gland (hypophysis)

Pituitary glands rests in the sella turcica suspended from floor of 3rd ventricle by pituitary stalk and is in close relationship to optic chiasma.

The adenohypophysis is derived from the ectoderm lining the primitive mouth cavity, and the neurohypophysis from an outgrowth of the hypothalamus.

Hypothalamic-hypophysial portal circulation forms functional link between hypothalamus and anterior pituitary.

HISTOLOGY

Cell types classified according to arbitrary staining techniques:

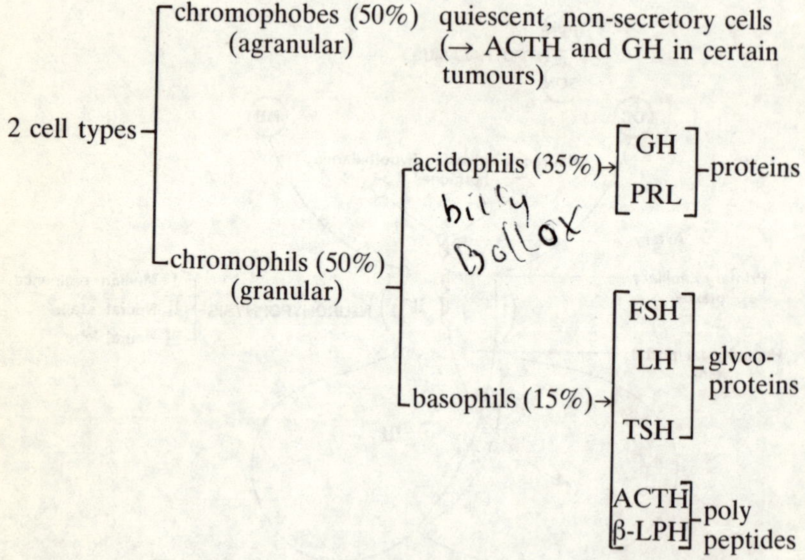

Ultrastructural and immunofluorescent studies suggest that each hormone is secreted by a specific cell type:

somatotrophs → GH

lactotrophs → PRL

FSH gonadotrophs → FSH (both hormones are probably secreted

LH gonadotrophs → LH by the same cells

thyrotrophs → TSH

*corticotrophs → ACTH, β-LPH (and related hormones in some species)

* A single cell type, in the anterior pituitary (in man) appears to secrete both ACTH and β-LPH

melanotrophs → MSH (not in man); these are present in pars intermedia

Lactotrophs are also known as mammotrophs.

HORMONES

1. GH or STH (growth hormone; somatotrophin) protein, MW 21 500 (hGH, human), 191 amino acids.
2. PRL or LTH (prolactin; lactogenic hormone; luteotrophic hormone) protein, MW 23 000 (ovine), 198 amino acids.
3. FSH (follicle-stimulating hormone) glycoprotein, MW 25 000 to 30 000 (hFSH, human).
4. LH (luteinising hormone); ICSH (interstitial cell-stimulating hormone) glycoprotein, MW 26 000 (hLH, human).
5. TSH (thyroid-stimulating hormone; thyrotrophin) glycoprotein, MW ~ 28 000 (hTSH, human), 209 amino acids.
6. ACTH (adrenocorticotrophic hormone) polypeptide, MW 4541 (hACTH, human), 39 amino acids.
7. LPH (lipotrophic hormone) polypeptide, MW ~ 9500 (human β-LPH), 91 amino acids, physiological role is unknown; a precursor of endogenous opiates (see p. 12).
8. MSH (melanocyte-stimulating hormone) polypeptide, α and β-MSH (see p. 12).

GH, PRL and HPL (human placental lactogen or human chorionic somatomammotrophin) have close structural affinities.

Hormone biosynthesis occurs in ribosomes of rough endoplasmic reticulum; granules are packaged by Golgi apparatus and then secreted by reverse pinocytosis into the blood. Lysosomal enzymes destroy unwanted granules.

The glycoprotein hormones (FSH, LH, TSH and placental HCG) have similar α-subunits and different β-subunits — these impart hormonal specificity. The α and β subunits of FSH and LH have been detected in plasma in pregnancy and in postmenopausal women, but their biological significance is unknown.

HYPOTHALAMUS-ANTERIOR PITUITARY RELATIONSHIPS

Until recently it was thought that specific areas in the hypothalamus modulate the release of specific anterior pituitary hormones, but the specificity of these relationships in man is now being questioned.

The nervous pathways into these hypothalamic 'centres' have not been clearly defined, and the chemical mediators at the afferent synapses may be any of a number of substances known to be present locally, viz. acetylcholine (Ach), noradrenaline (NA), dopamine (DA), serotonin (5HT), substance P, and histamine (sometimes called monoaminergic neurones).

The 'centres' consist of diffuse overlapping networks of neurones (sometimes called peptidergic neurones) and each 'centre' regulates the secretion of one or more anterior pituitary hormone by the release of a hypothalamic hormone (neurohormone) into the primary capillary plexus in the median eminence. The releasing factor (or inhibitory factor) is conveyed to the anterior pituitary via the portal system of vessels where it influences the secretion of one or more hormones (see Fig. 1.2).

The postulated hypothalamic (hypophysiotrophic) hormones are:

GRF (GH-releasing factor)

GIF (GH-inhibitory factor; somatostatin)

PIF (PRL-inhibitory factor; dopamine)

PRF (PRL-releasing factor)

FRF (FSH-releasing factor) }
LRF (LH-releasing factor) } a single substance, gonadotrophin releasing hormone; GRH (or GnRH)

TRF (TSH-releasing factor)

CRF (Corticotrophin-(ACTH-) releasing factor)

Synonyms: 'factor' and 'hormone' are now commonly used interchangeably in these names: generally called 'hormone' if chemical structure has been characterised (e.g. TRH, LRH, GIH). A systematic nomenclature is being introduced in which releasing factors have names ending in '-liberin' (e.g. TRF = thyroliberin) and inhibitory factors, in '-statin' (e.g. GIF = somatostatin). Note that this nomenclature applies to hypothalamic hormones, cf. pituitary hormones' systematic names end in '-trophin' (e.g. TSH = thyrotrophin).

CNS impulses $\xrightarrow[\text{(or)}-]{+}$ monoaminergic neurone $\xrightarrow[\text{(or)}-]{+}$ peptidergic neurone $\xrightarrow[\text{(or)}-]{+}$ hypophysiotrophic hormone/factor $\xrightarrow[\text{(or)}-]{+}$ anterior pituitary hormone secretion. The sensitivity of anterior pituitary cell to hypophysiotrophic peptide modulates its rate of secretion.

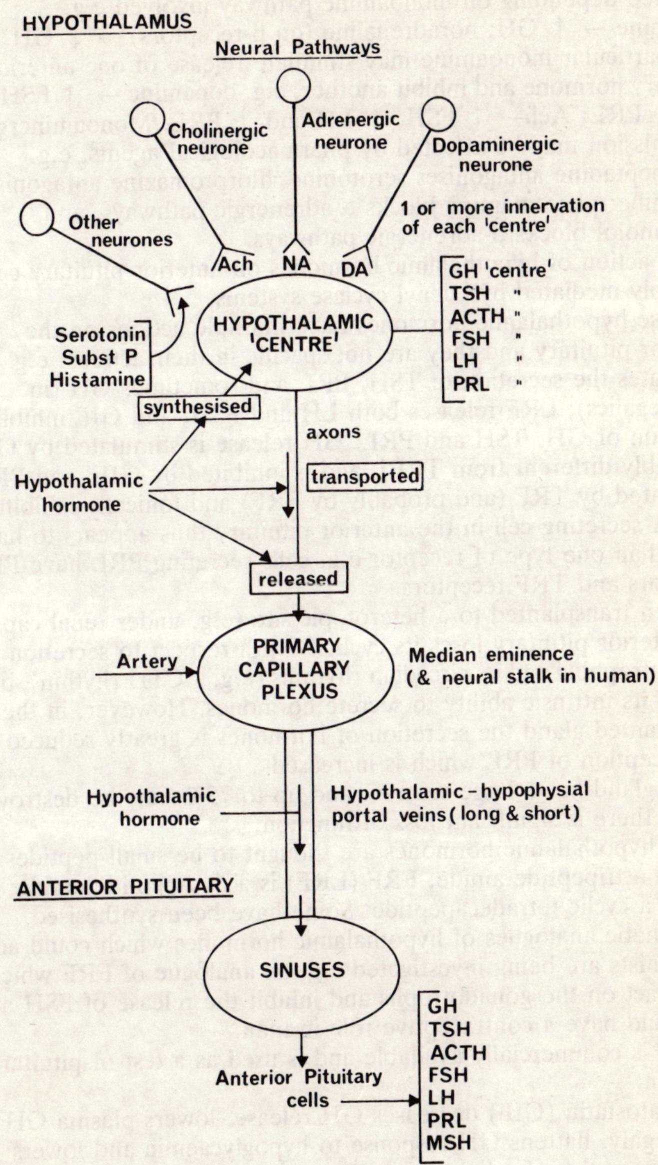

Fig. 1.2 Hypothalamus-anterior pituitary relationship

A particular anterior pituitary hormone may be stimulated or inhibited depending on monoamine pathway involved e.g. dopamine → ↑ GH; noradrenaline (on β-receptors) → ↓ GH.

A particular monoamine may stimulate release of one anterior pituitary hormone and inhibit another, e.g. dopamine → ↑ FSH and ↓ PRL; Ach→ ↑ FSH, ↑ LH and ↓ PRL. Monoaminergic transmission may be affected by pharmacological agents, e.g. cyproheptadine antagonises serotonin; chlorpromazine antagonises dopamine; phentolamine blocks α-adrenergic pathways, and propranolol blocks β-adrenergic pathways.

The action of hypothalamic hormones on anterior pituitary cells is probably mediated by adenyl cyclase systems.

These hypothalamic hormones have multiple actions on the anterior pituitary and they are not specific in their actions, e.g. TRF stimulates the secretion of TSH, PRL and sometimes GH (in acromegalics); LRF releases both LH and FSH; and GIF inhibits secretion of GH, TSH and PRL. GH release is stimulated by GRF (probably different from TRF), and is inhibited by GIF; and PRL is stimulated by TRF (and probably by PRF) and tonically inhibited by PIF. A secreting cell in the anterior pituitary thus appears to have more than one type of receptor e.g. cells secreting PRL have PIF receptors and TRF receptors.

When transplanted to a heterotopic site (e.g. under renal capsule) the anterior pituitary loses its cyclicity with respect to secretion of gonadotrophins and its circadian rhythms (e.g. ACTH rhythm), but retains its intrinsic ability to secrete hormones. However, in the transplanted gland the secretion of hormones is greatly reduced with the exception of PRL which is increased.

The gland has a large reserve, and up to 75% may be destroyed before there is significant loss of function.

The hypothalamic hormones are thought to be small peptides: TRF is a tripeptide amide; FRF (LRF) is a decapeptide amide; and GIF is a cyclic tetradecapeptide. Some have been synthesised.

Synthetic analogues of hypothalamic hormones which could act as antagonists are being investigated e.g. an analogue of FRF which would act on the gonadotrophs and inhibit the release of FSH and LH could have a contraceptive role in man.

TRF is commercially available and is used as a test of pituitary function.

Somatostatin (GIF) decreases GH release, lowers plasma GH in acromegaly, flattens GH response to hypoglycaemia and lowers insulin secretion. In diabetes mellitus, somatostatin decreases plasma glucose by lowering glucagon secretion.

Hypothalamus-anterior pituitary target gland servo-system
This is illustrated in Figure 1.3.

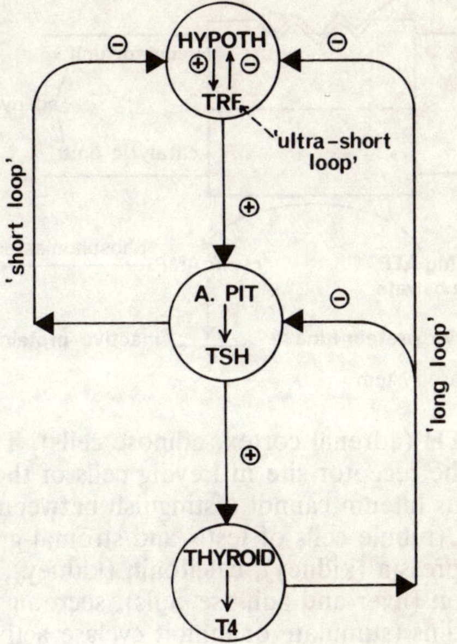

Fig. 1.3 Hypothalamus-anterior pituitary target gland servo-system, e.g. control of thyroid secretion

Target gland hormone exerts negative feedback control over production of anterior pituitary hormone by direct action on anterior pituitary and indirectly through hypothalamus. The degree of this 'longloop' regulation varies with hormone systems. 'Short loop'and 'ultra-short loop' servo-systems are also thought to exist, but the mechanisms are not clear.

Mechanisms of hormone action
The first step in the modulation of tissue activity is the combination of hormone with a receptor in or on the target cell. Receptors for protein, polypeptide hormones and neurotransmitters are located at the cell membrane. Receptors for steroid hormones are generally intracellular (though the receptor sites for glucocorticoids have not been located).

The adenyl cyclase system mediates the actions of many hormones (see Fig. 1.4). The hormonal specificity is conferred by specific receptor sites unique to each target tissue while the catalytic component of the system is very similar in all the target tissues. The following hormones are currently believed to exert their actions by

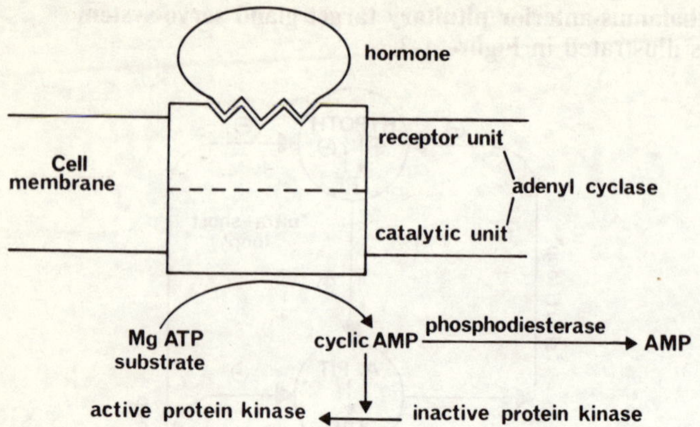

Fig. 1.4 Adenyl cyclase system

this system; ACTH (adrenal cortex, adipose cells), TSH (thyroid), LH and HCG (the receptor site in Leydig cells of the testis and in cells of the corpus luteum cannot distinguish between the two hormones), FSH (tubule cells of testis and stromal-granulosal cells of the ovary), vasopressin (kidney), calcitonin (kidney), parathormone (kidney), glucagon (liver and adipose cells), secretin (adipose cells), some prostaglandins (stimulate or inhibit cyclase activity in many tissues), adrenaline (β-adrenergic actions on heart, liver, adipose cells and probably other β-adrenergic actions).

The intracellular second messenger, cyclic AMP formed in response to stimulation of the adenyl cyclase activates protein kinases by binding to the repressor subunit of the kinase causing its dissociation and unmasking the active enzyme. Protein kinases catalyse the phosphorylation of metabolic enzymes, other protein kinases or genetic repressors, thus controlling biochemical activities.

The mechanisms of action of other hormones and neurotransmitters which act on the plasma membrane by mechanisms other than adenyl cyclase activation, are not known, but are thought in many cases to involve changes in membrane permeability. Sex hormones and mineralocorticoids (steroids) combine with intracellular receptors which control gene repression/expression.

Functions of growth hormone

Human growth hormone (hGH), MW 21 500, consists of 191 amino acids with 2 disulphide bridges, and has been synthesised. Species specific (only human and primate GH active in man; other species GH is metabolically inactive in man and evokes immune response). Several forms of GH are present in anterior pituitary and plasma: 'big' GH (dimeric or trimeric GH); 'little' GH (monomeric GH); GH-variant (MW ~ 20 000).Active fragments of GH have been

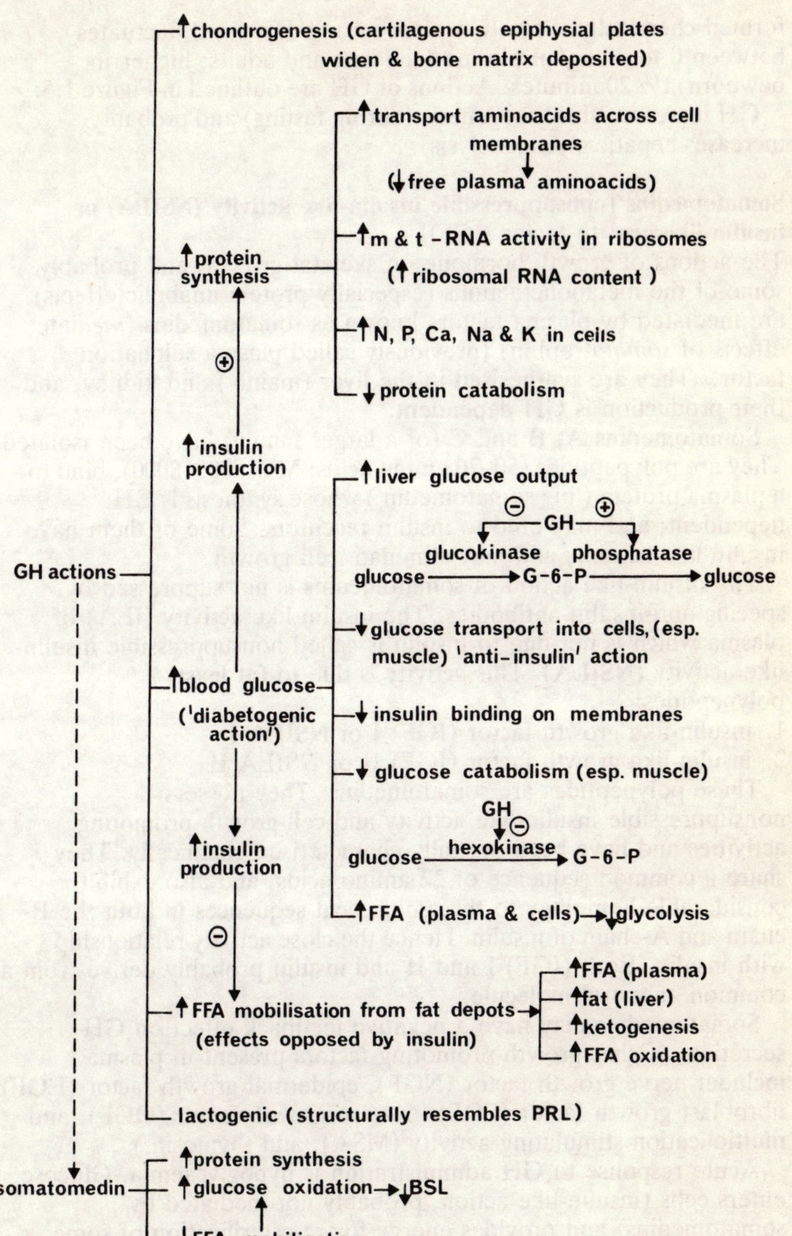

Fig. 1.5 Actions of growth hormone

formed chemically. Secretion rate ~ 1 mg/day; level fluctuates between 0 to 30 ng/ml plasma (children and adults; higher in newborn) t½ 20 minutes. Actions of GH are outlined in Figure 1.5.

GH increases gluconeogenesis (during fasting) and probably increases hepatic glycogenolysis.

Somatomedins [nonsuppressible insulin-like activity (NSILA) or insulin-like growth factor (IGF)]

The actions of growth hormone on skeletal growth, and probably some of the metabolic actions (especially protein anabolic effects), are mediated by plasma factors known as somatomedins (*med*iate effects of *somato*trophin) (previously called plasma sulphation factor). They are synthesised in the liver (mainly) and kidney, and their production is GH dependent.

Somatomedins A, B and C (of a larger family) have been isolated. They are polypeptides (50–70 amino acids; MW 5000–8000); bind to a plasma protein ('big somatomedin') whose synthesis is GH dependent; and may bind to insulin receptors. Some of them have insulin-like actions, and may stimulate cell growth.

The insulin-like action of somatomedins is not suppressed by specific anti-insulin antibodies. The insulin-like activity (ILA) of plasma which is not due to insulin is called nonsuppressible insulin-like activity (NSILA). This activity is due to (at least) 2 polypeptides:

1. insulin-like growth factor (IGF) I or NSILA I
2. insulin-like growth factor (IGF) II or NSILA II

These polypeptides are somatomedins. They possess nonsuppressible insulin-like activity and cell-growth-promoting activities and have been partially characterised chemically. They share a common sequence of 22 amino acids, and also exhibit considerable homology to the amino acid sequences in both the B-chain and A-chain of insulin. Hence the close activity relationship with insulin. Both (IGF) I and II and insulin probably derive from a common ancestral molecule.

Somatomedins may have a negative feedback effect on GH secretion. (Other growth-promoting factors present in plasma include: nerve growth factor (NGF), epidermal growth factor (EGF), fibroblast growth factor (FGF), ovarian growth factor (OGF), and multiplication-stimulating activity (MSA), and thymosin.)

Acute response to GH administration is hypoglycaemia. Glucose enters cells (insulin-like action, probably not mediated by somatomedins) and provides energy for re-esterification of some FFA following lypolytic effect of GH.

In Laron type dwarfism, plasma GH is high and plasma somatomedin is low. GH therapy does not increase somatomedin.

African pygmies and Turner's syndrome dwarves have normal plasma levels of GH and somatomedin, but the peripheral tissues are nonresponsive.

Growth hormone stimulates the β-cells to secrete insulin; in chronic doses it causes degenerative changes in these cells and decreased insulin production (metahypophysial diabetes). In some cases of acromegalic diabetes, the level of plasma insulin is raised and the binding of insulin peripherally is reduced (i.e. true insulin resistance).

Growth hormone-insulin relationship
This is shown in Figure 1.6.

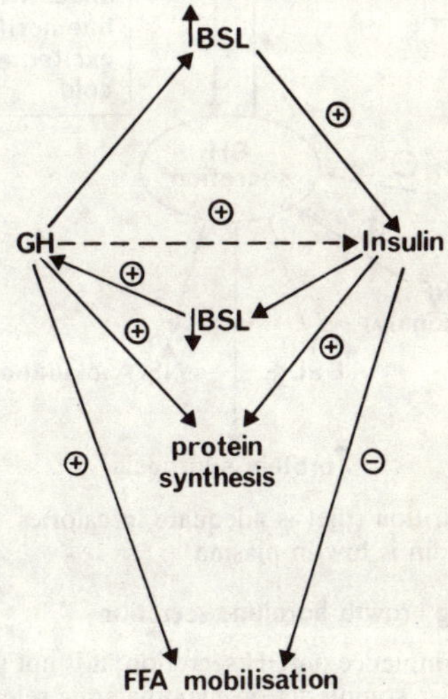

Fig. 1.6 Growth hormone-insulin relationship

Under physiological conditions these actions of GH tend to be balanced by secretion of insulin.

Factors influencing growth hormone secretion
See Figure 1.7.

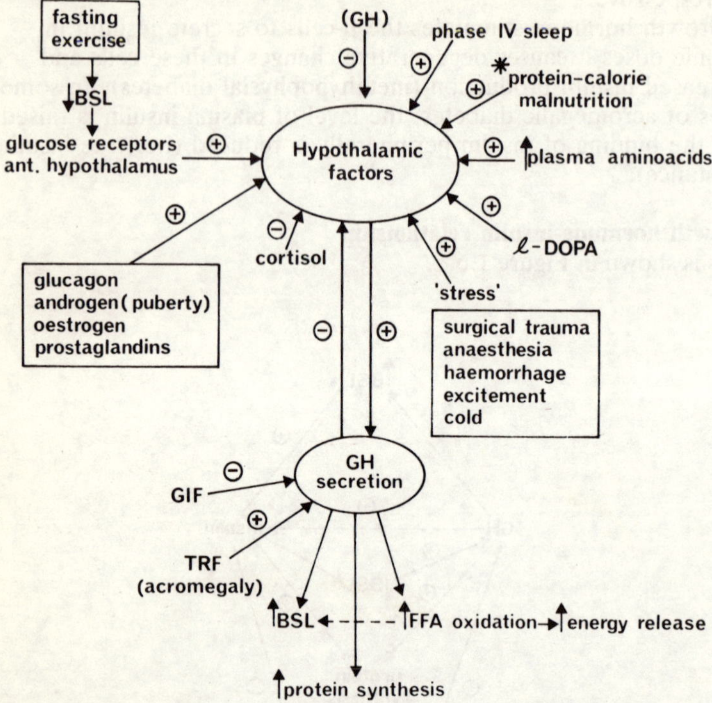

* In protein-calorie malnutrition (diet is adequate in calories, but deficient in protein) GH is high and somatomedin is low in plasma

Fig. 1.7 Factors influencing growth hormone secretion

Key: ⊕ and ⊖ refer to the influence on GH secretion; it is not possible at present to determine whether particular stimuli act on hypothalamic releasing or inhibiting factors; almost all influences are mediated by the adrenergic system (α-adrenergic pathways stimulate, and β-adrenergic pathways inhibit, GH release)
BSL = blood glucose concentration

ACTH-related peptides
These include: β-lipotrophin (β-LPH), α- and β-melanocyte-stimulating hormones (MSH), γ-lipotrophin (γ-LPH), corticotrophin-like intermediate lobe peptide (CLIP), endorphins (α, β and γ), (methionine) Met-enkephalin and (leucine) Leu-enkephalin.

ACTH and β-LPH are linear peptides with 39 and 91 amino acids, respectively. They are formed from a large precursor hormone (MW 31 000) in the same basophil cells in the pars distalis and stored in the same granules. Both ACTH and β-LPH undergo selective cleavage to produce peptide fragments with certain common amino acids sequences.

These related peptides with the amino acid sequence of the parent hormone in brackets are:

α-MSH = ACTH-(1–13)

CLIP = ACTH-(18–39)

γ-LPH = β-LPH-(1–58)

β-MSH = β-LPH-(37–58)

β-endorphin = β-LPH-(61–91)

γ-endorphin = β-LPH-(61–77)

α-endorphin = β-LPH-(61–76)

Met-enkephalin = β-LPH-(61–65)

Intermediate lobe hormones
In species with well developed intermediate lobes (not in man): a large precursor molecule → ACTH and β-LPH; ACTH → α-MSH and CLIP; β-LPH → γ-LPH which is converted to β-MSH; α- and β-MSH and CLIP are secreted into the blood; and ACTH, β-LPH and γ-LPH are retained within the gland.

In man (with rudimentary intermediate lobes): a large precursor molecule → ACTH, β-LPH and β-endorphin; these are secreted without further fragmentation into the blood. Hence, since α- and β-MSH, CLIP and γ-LPH are not formed in the gland they are not secreted into the blood. The MSH-like activity of human blood is due to ACTH (largely) and β-LPH which possess MSH-like activity. The secretion of both ACTH and β-LPH is in parallel; is influenced by the same stimuli; and is mediated by CRF.

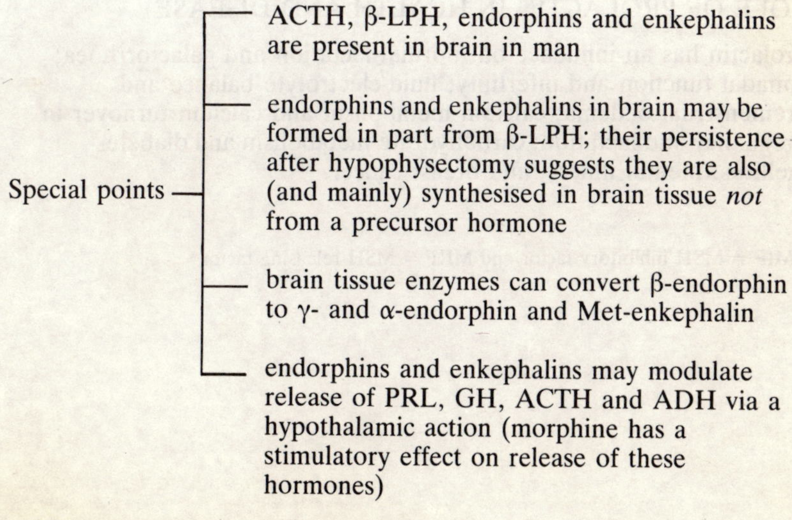

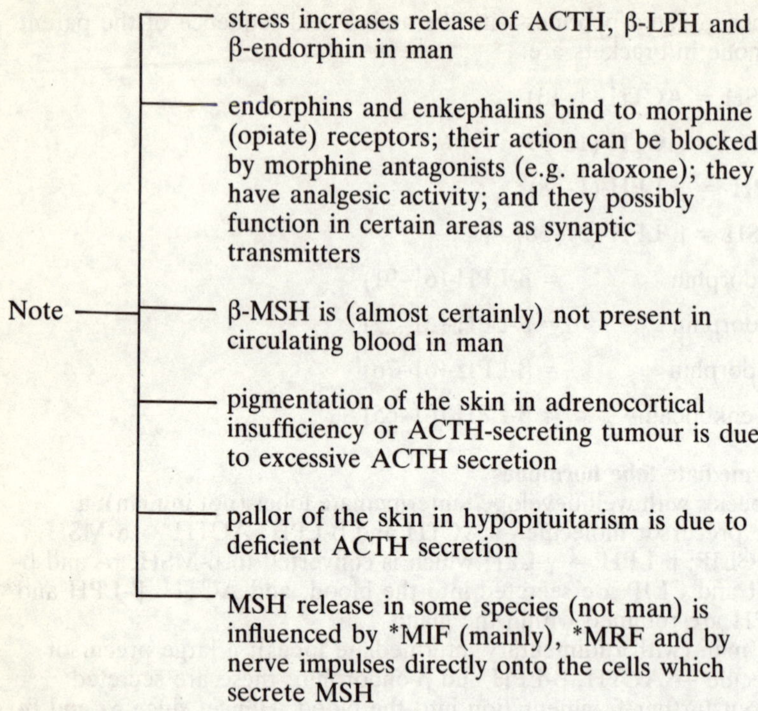

Other anterior pituitary hormones

These are considered in the description of the gland on which they act.

ROLE OF PROLACTIN IN HEALTH AND DISEASE

Prolactin has an influence on: normal lactation and galactorrhoea; gonadal function and infertility; fluid/electrolyte balance and premenstrual oedema; calcium metabolism and calcium turnover in pregnancy and lactation; carbohydrate metabolism and diabetes mellitus; mental illness; and breast cancer.

* MIF = MSH inhibitory factor and MRF = MSH releasing factor

Regulation of prolactin secretion
See Figure 1.8.

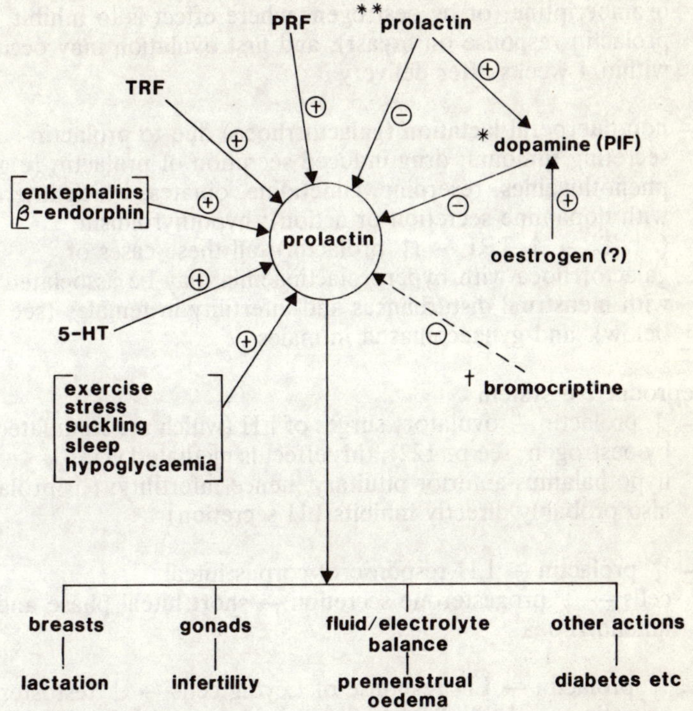

* = major controlling factor
** = at pituitary level, prolactin probably inhibits its own secretion by enhancing effect of dopamine on prolactin-secreting cells; at hypothalamic level, prolactin stimulates secretion of dopamine
† = bromocriptine (2-bromo-α-ergocryptine) is a lysergic acid derivative with agonist effects on postsynaptic dopaminergic receptors in hypothalamus

Fig. 1.8 Regulation of prolactin secretion

Note: enkephalins and β-endorphin stimulate secretion of prolactin and inhibit secretion of FSH/LH, hence they may affect sexual function

Actions of prolactin

Prolactin affects the following:

1. Lactation
   - puerperal lactation (see p. 222) may be suppressed by bromocriptine (or by oestrogens where effect is to inhibit prolactin response on breast), and first ovulation may occur within 4 weeks after delivery
   - non-puerperal lactation (galactorrhoea) due to prolactin-secreting tumours; drug induced secretion of prolactin (e.g. phenothiazines, reserpine, cimetidine, opiates, which interfere with dopamine secretion or action); hypothyroidism (↓ T_4 → ↑ TRF → ↑ prolactin): all these cases of galactorrhoea with hyperprolactinaemia may be associated with menstrual disturbances and infertility in females (see below), and gynaecomastia in males

2. Reproductive system
   - ↑ prolactin $\xrightarrow{-}$ ovulatory surges of LH (which are stimulated by oestrogen, see p. 122); this effect is mediated via hypothalamus-anterior pituitary, hence infertility; (↑ prolactin also probably directly inhibits LH secretion)
   - ↑ prolactin $\xrightarrow{-}$ LH response of corpus luteal cells → ↓ progesterone secretion → short luteal phase and amenorrhoea
   - ↑ prolactin $\xrightarrow{-}$ LH response of Leydig cells → ↓ testosterone secretion → ↓ libido, erectile impotence and infertility; prolactin also reduces conversion of testosterone to its active metabolite dihydrotestosterone

3. Fluid and electrolyte metabolism

   ↑ prolactin (in luteal phase) → ↑ renal retention of Na, K, H_2O → fluid retention → premenstrual oedema (steroids, e.g. oestrogen, and ADH may also contribute to oedema formation in this syndrome), bromocriptine therapy may be successful

4. Calcium metabolism

   prolactin increases conversion of 25 hydroxycholecalciferol to 1,25 dihydroxycholecalciferol in renal tubules (see p. 201), and thus has an important role on calcium metabolism during pregnancy and lactation

5. Carbohydrate metabolism
   prolactin has similar actions to GH (see p. 9) hence, ↑ prolactin → ↓ glucose tolerance and ↑ insulin secretion, and thus is diabetogenic

6. Mental illness
   prolactin increases PGE_1 secretion and, thus, increases the secretion of PGE_1 which is reduced in schizophrenia (note: antischizophrenic drugs e.g. phenothiazines and butyrophenones are dopamine antagonists and, hence, increase prolactin secretion)

7. Breast cancer
   prolactin (and GH) increase growth of breast cancer; hence, suppression of both hormones is effective in some prolactin/GH dependent breast cancers

Note: bromocriptine has an agonist effect on postsynaptic dopamine receptors in the nigrostriatum and possibly limbic system and, hence, has an anti-Parkinsonian action.

Disorders of anterior pituitary function

Hypopituitarism
This may be primary, when the lesion is in the anterior pituitary or secondary, when the lesion involves the hypothalamus and the synthesis/release of hypothalamic (releasing/inhibitory) factors (hormones).

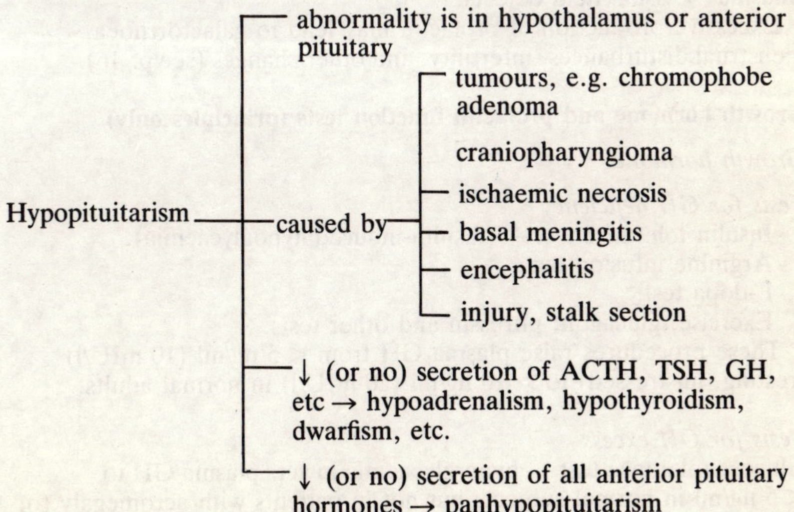

Liability to coma is a feature of panhypopituitarism; and this may be due to hypoglycaemia in patients with increased sensitivity to insulin, to water intoxication in patients with adrenal insufficiency, to severe hypothermia in patients with hypothyroidism, or to disorders of electrolyte and acid-base balance.

Hyperpituitarism
Hyperfunction of the anterior pituitary may be primary, when the lesion is in the anterior pituitary or secondary, when the lesion involves the hypothalamus and the synthesis/release of hypothalamic factors.

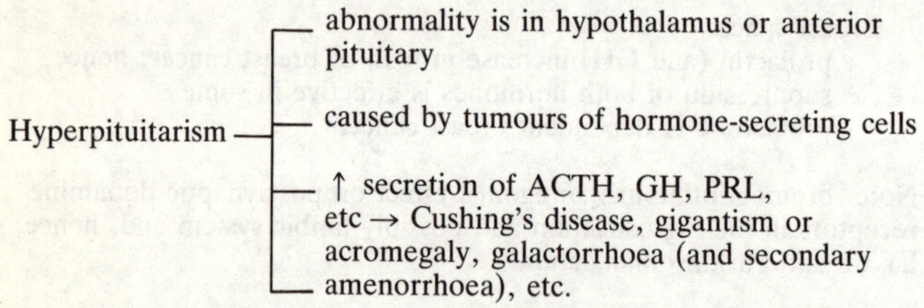

Excessive production of GH before linear bone growth ceases → gigantism, and after epiphysial fusion → acromegaly. Both conditions may be associated with splanchnomegaly, hyperglycaemia, reduced glucose tolerance, and a tendency to develop diabetes mellitus. Destruction of the anterior pituitary and, hence, hypopituitarism may occur. Compression of the optic chiasma → visual field defects.

Excessive production of prolactin may lead to galactorrhoea, menstrual disturbances, infertility, and other changes (see p. 16).

Growth hormone and prolactin function tests (principles only)

Growth hormone

Tests for GH deficiency
1. Insulin tolerance test — (insulin-induced hypoglycaemia).
2. Arginine infusion test.
3. L-dopa test.
4. Exercise; glucagon; pitressin and other tests.

These procedures raise plasma GH from < 5 ng/ml (10 mIU/l) (resting, unstressed) to > 10 ng/ml (20 mIU/l) in normal adults.

Tests for GH excess
Glucose tolerance test — hyperglycaemia lowers plasma GH to < 5 ng/ml in normal subjects but not in patients with acromegaly (or pituitary gigantism).

Prolactin

1. Measurement of prolactin — normal resting plasma PRL values are 5 to 25 ng/ml (60–300 mlU/1), and these are raised in patients with PRL-secreting pituitary tumours.
2. Chlopromazine stimulation test — raises plasma PRL in normal subjects but not in hypopituitarism.
3. Metoclopramide stimulation test — raises plasma PRL in normal subjects but less so in hypopituitarism.
4. Insulin tolerance test — same response as (2) above.
5. Bromergocryptine (bromocriptine) and L-dopa — lower plasma PRL in normal subjects and in patients with PRL-secreting tumours.

FURTHER READING

Besser, G. M. (ed.) (1977) The hypothalamus and pituitary. *Clinical Endocrinology*, **6**, 1.

Bloom, F. E. (1979) Contrasting principles of synaptic physiology: Peptidergic and non-peptidergic neurons. In Fuxe, K., Hökfelt, T., Luft, R. (eds) *Central Regulation of the Endocrine System*, p. 173–187. New York: Plenum Press.

Chan, J. S. D., Lu, C-L., Seidah, N. G., Chretien, M. (1982) Corticotropin releasing factor (CRF) : effects on the release of pro-opiomelanocortin (POMC)-related peptides by human anterior pituitary cells in vitro. *Endocrinology*, **111**, 1388 (rapid communication).

Collu, R., Barbeau, A., Ducharme, J. R., Rochefort, J-G. (eds) (1979) *Central Nervous System Effects of Hypothalamic Hormones and other Peptides*. New York: Raven Press.

Costa, E., Trabucchi, M. (eds) (1977) Endorphins. *Advances in Biochemical Psychopharmacology 18*. New York: Raven Press.

Daughaday, W. H. (1977) Hormonal regulation of growth by somatomedin and other tissue growth factors. *Clinical Endocrinology and Metabolism*, **6**, 117.

De Souza, E. B., Van Loon, G. R. (1982) D-Ala^2-Met-Encephalinamide, a potent opioid peptide, alters pituitary-adrenocortical secretion in rats. *Endocrinology*, **111**, 1483.

Elde, R., Hökfelt, T. (1979) Localization of hypophysiotrophic peptides and other biologically active peptides within the brain. *Annual Review of Physiology*, **41**, 587.

Fuxe, K., Hökfelt, T., Luft, R. (eds) (1979) *Central Regulation of the Endocrine System*. New York: Plenum Press.

Frantz, A. G. (1978) Prolactin. *New England Journal of Medicine*, **298**, 201.

Herman, M. L., Ben-Jonathan, N. (1982) Rat anterior pituitary dopaminergic receptors are regulated by estradiol during lactation. *Endocrinology*, **111**, 1057.

Jeffcoate, S. L., Hutchinson, J. S. M. (1978) *The Endocrine Hypothalamus*. London: Academic Press.

Johns, M. A., Azmitia, E. C., Krieger, D. T. (1982) Specific *in vitro* uptake of secrotonin by cells in the anterior pituitary of the rat. *Endocrinology*, **110**, 754–760.

Kosterlitz, H. W. (1978) Endogenous opioid peptides: historical aspects. In Hughes J. (ed.) *Centrally Acting Peptides*, p. 157. Baltimore: University Park Press.

Krulich, L. (1979) Central neurotransmitters and the secretion of prolactin, GH, LH and TSH. *Annual Review of Physiology*, **41**, 603.

Martini, I., Ganong, W. F. (eds) (1978) *Frontiers in Neuroendocrinology*, vol. 5. New York: Raven Press.

Morris, D. H., Sehalch, D. S. (1982) Structure of somatomedin-binding protein : alkaline pH-induced dissociation of an acid-stable, 60 000 molecular weight complex into smaller components. *Endocrinology*, **111**, 801.

Motto, M., Crossignani, P. G., Martini, L. (eds) (1975) Hypothalamic hormones: chemistry, physiology, pharmacology and clinical uses. *Proceedings of Sereno Foundation Symposium No. 6*, p. 350. London: Academic Press.

Polonsky, K., Jaspan, J., Berelowitz, M., Pugh, W., Moossa, A., Ling, N. (1982) The *in vivo* metabolism of somatostatin 28: possible relationship between diminished metabolism and enhanced biological action. *Endocrinology*, **111**, 1698.

Van Wyk, J. J., Underwood, I. E. (1978) The somatomedins and their actions. In Litwack, G. (ed.) *Biochemical Actions of Hormones*, vol. V, p. 102–148. New York: Academic Press.

Watkins, W. B. (1977) *Hypothalamic Releasing Factors*, vol. I. Edinburgh: Churchill Livingstone.

Watkins, W. B. (1978) *Hypothalamic Releasing Factors*, vol. 2. Edinburgh: Churchill Livingstone.

Multiple choice questions

1. Insulin-like growth factor (IGF):
1. is a somatomedin;
2. has nonsuppressible insulin-like activity (NSILA);
3. has close structure/activity relationship to insulin;
4. is a steroid;
5. has close structure/activity relationship to GH;
6. level in plasma is GH-dependent.

2. Somatomedins:
1. are polypeptides;
2. may have insulin-like activity;
3. are carried in plasma bound to a GH-dependent protein;
4. may compete with insulin for common receptors;
5. are established hypophysiotrophic factors;
6. have no structure/activity relationship to NSILA I;
7. are present in high concentration in plasma in acromegalics.

3. ACTH-related peptides include:
1. α-MSH, which is present in circulating blood in man;
2. CLIP, which is derived from β-LPH in man;
3. β-endorphin, which can stimulate GH secretion;
4. Met-enkephalin, which is secreted by anterior pituitary in man;
5. endorphins and enkephalins, which bind to morphine receptors.

4. The effects of GH on protein metabolism are to:
1. decrease blood amino acid concentration;
2. increase urinary urea-nitrogen;
3. increase ribosomal RNA synthesis;
4. induce a positive nitrogen balance;
5. increase protein synthesis by an action mediated wholly via insulin.

5. With respect to ACTH-related peptides:
1. hyperpigmentation of the skin in adrenocortical insufficiency in man results from excessive production of β-MSH;
2. pallor of the skin in hypopituitarism in man results from deficient ACTH secretion;
3. Met-enkephalins may act as neurotransmitters in the brain
4. they are produced in the adrenal cortex;
5. endorphins can stimulate ADH secretion.

6. The following hormones are secreted by the acidophilic cells of the adenohypophysis:
1. adrenocorticotrophic hormone (ACTH);
2. growth hormone (GH);
3. thyroid-stimulating hormone (TSH);
4. luteinising hormone (LH);
5. follicle-stimulating hormone (FSH);
6. prolactin (PRL).

7. Growth hormone deficiency in humans:
1. has biochemical, but no clinical effects in adults;
2. is responsible for all cases of short stature;
3. can be corrected by bovine and ovine growth hormone injections;
4. can be corrected by injections of human anterior pituitary extracts;
5. can be recognised by a deficient growth hormone response to exercise;
6. can be recognised by a deficient growth hormone response to hyperglycaemia during a glucose tolerance test.

8. When the anterior pituitary is transplanted to a heterotopic site (e.g. under the renal capsule):
1. it cyclically releases LH;
2. the circadian rhythm of ACTH secretion is retained;
3. the secretion of FSH is markedly decreased;
4. the secretion of PRL is markedly increased;
5. the secretion of TSH ceases;
6. it is unresponsive to circulating endogenous releasing factors because of their minute concentration in the systemic blood;
7. the synthesis of CRF in the hypothalamus is increased;
8. the synthesis of oxytocin in the paraventricular nuclei of the hypothalamus is decreased.

9. Chromophobe adenomas of the pituitary:
1. can be associated with acromegaly;
2. can be associated with galactorrhoea;
3. are non-functional;
4. always have some clinical effect;
5. may present with visual field impairment;
6. may cause symptomless pituitary fossa enlargement.

10. With respect to anterior pituitary hormones in the human:
1. both ACTH and β-LPH are split from a large precursor molecule;
2. the secretion of ACTH parallels that of β-LPH;
3. MSH is presumably not secreted into the blood;

4. MSH-like activity of human blood is presumably due mainly to ACTH;
5. some endorphins and enkephalins are possibly derived from β-LPH.

11. Growth hormone levels in acromegaly:
1. are increased in sleep;
2. are increased by L-dopa;
3. are suppressed by bromergocryptine;
4. are increased by glucose;
5. are increased by TRF;
6. must fall to normal if acral regression is to occur.

12. With respect to hypothalamic hormones (i.e. releasing or inhibitory factors):
1. castration increases the LRF content in the hypothalamus thus suggesting a long-loop feedback mechanism;
2. GIF has been synthesised and is a 14 amino acid peptide;
3. TRF injected into cretins can induce TSH secretion;
4. GIF is somatostatin and it lowers insulin secretion;
5. somatostatin increases glucagon secretion;
6. TRF stimulates secretion of TSH and has no effect on PRL;
7. their action on anterior pituitary cells is mediated by adenyl cyclase systems.

13. The following statements refer to patients who have undergone hypophysectomy:
1. the rate of glucose absorption from the gut is increased;
2. the rate of gluconeogenesis is decreased;
3. insulin sensitivity is increased;
4. if diabetes insipidus co-exists, cortisol administration can exacerbate the condition;
5. in diabetes mellitus, the diabetic state becomes more severe;
6. mineralocorticoid replacement therapy is usually (> 50% of cases) required.

14. Growth hormone:
1. increases the concentration of hormone sensitive lipase in fat cells;
2. secretion is increased during acute haemorrhage (e.g. 1l blood loss in 1 h) and the hyperglycaemia produced helps to expand the blood volume;
3. occurs in the plasma of the adult man around 0 to 2 ng/ml (postprandially);
4. is synthesised in polysomes of the rough, endoplasmic reticulum and is packaged into granules in the Golgi complex;

5. secretion increases during sleep, and this response helps to maintain a 'steady' blood glucose level which is physiologically important because glucose is the major energy substrate for nerve cells;
6. decreases the mobilisation of FFA from fat cells;
7. decreases RNA-ribosome content in muscle cells;
8. decreases the transport of amino acids across muscle cell membranes;
9. secretion is induced by hypoglycaemia and this response tends to maintain plasma glucose homeostasis.

15. With regard to human pituitary prolactin (PRL):
1. TRF administration is followed by a rise in serum PRL levels;
2. during the puerperium, serum PRL levels are elevated while gonadotrophins are low;
3. some ergot derivatives are useful clinically in suppressing pituitary PRL secretion;
4. some phenothiazine derivatives acutely lower serum PRL concentration;
5. dopamine is considered to be the major biogenic amine involved in the stimulation of PRL secretion;
6. chromophobe adenomata quite commonly (> 20%) secrete PRL;
7. induced falls in serum osmolality (e.g. by administration of a water load) cause serum PRL levels to fall.

16. With respect to hypothalamic hormones (i.e. releasing or inhibitory factors):
1. FRF and LRF are the same decapeptide amide;
2. TRF is a tripeptide amide and is commercially available;
3. GIF inhibits secretion of GH and has no effect on TSH;
4. mammotrophs have receptors for PIF and no receptors for any other hypothalamic hormone;
5. FSH administration decreases FRF content in the hypothalamus of castrated animals thus suggesting a short-loop feedback mechanism;
6. they are produced by discrete anatomical centres in the hypothalamus;
7. FRF injected into anovulatory women can induce FSH secretion.

17. Growth hormone:
1. decreases the activity of lipoprotein lipase in the vicinity of fat cells;
2. secretion is increased during starvation (e.g. 48 h) and this response increases the production of glucose and ketone bodies by the liver;

3. has effects on growth of duct and ductule-alveolar systems in the breast and on the maintenance of milk secretion;
4. is a glycoprotein;
5. production in cultured anterior pituitary tissue is increased by hypothalamic extracts which contain GRF;
6. increases protein synthesis mainly because it causes secretion of insulin;
7. decreases liver glucogenesis;
8. increases the reaction glucose → G-6-P in liver cells by increasing the concentration of glucokinase.

18. Acromegaly is an endocrine disorder in which:
1. radiological evidence of pituitary enlargement is invariable;
2. growth hormone levels are elevated but somatomedin levels are low;
3. visceromegaly as well as bony overgrowth occur;
4. headache and sweating are common complaints;
5. impotence, when it occurs, is due to gonadotrophin deficiency;
6. abnormal responses of growth hormone to physiological stimuli occur;
7. patients frequently die from complications of diabetes mellitus;
8. patients may require treatment for cardiac failure.

19. FRF (i.e. LRF) differs from FSH in that it:
1. is present in the hypothalamus;
2. is a polypeptide;
3. is present in the systemic circulation;
4. is secreted during the menstrual cycle in amounts which vary with the plasma oestrogen level;
5. is produced in greater quantity when the anterior pituitary is transplanted to an ectopic site;
6. is subject to negative feedback control by the plasma oestrogen level during the menstrual cycle;
7. is secreted in man in decreased amounts when testosterone is administered in adequate dosage.

20. The following statements refer to growth hormone secretion:
1. fasting growth hormone levels provide a useful test to differentiate panhypopituitarism from normal;
2. muscular exercise causes a rise in plasma growth hormone levels;
3. no significant changes in plasma growth hormone levels are seen at night;
4. the single most useful test in the diagnosis of acromegaly is the response of plasma growth hormone to glucose administration;

5. growth hormone administration is followed acutely by a fall in the level of blood glucose;
6. cortisol opposes the growth promoting actions of growth hormone;
7. children with growth hormone deficiency characteristically have no change in their skeletal age.

21. Somatomedin:
1. mediates the protein anabolic effects of GH;
2. is low in plasma in kwashiorkor although GH is high;
3. is the growth hormone secreted by the pars intermedia;
4. when absent in plasma may be associated with dwarfism although GH in these patients is high;
5. is a powerful myocardial stimulant;
6. has some insulin-like actions;
7. competes with insulin receptors in some cells, e.g. in liver and adipose tissue cells;
8. in some cells acts on receptors which are distinct from insulin receptors.

22. Hypopituitarism:
1. can be due to a pituitary tumour as well as atrophy of the gland;
2. may be of sudden or gradual onset;
3. is generally associated with increased skin pigmentation;
4. may not involve deficiency of all pituitary hormones;
5. leads to insulin sensitivity and hypoglycaemia;
6. requires special care during general anaesthesia.

23. Galactorrhoea:
1. invariably means a pituitary tumour is present;
2. is generally associated with increased prolactin levels;
3. is a complication of oestrogen therapy;
4. is a complication of phenothiazine therapy;
5. is a complication of corticosteroid therapy;
6. never occurs in males;
7. is improved by L-dopa therapy;
8. may follow pituitary stalk section.

24. With respect to mechanism of hormone action: which is the odd one out?
1. vasopressin;
2. TSH;
3. FSH;
4. HCG;
5. testosterone;
6. ACTH;
7. adrenaline.

25. With respect to the kidney: which is the odd one out?
1. adrenaline;
2. PTH;
3. calcitonin;
4. vasopressin;
5. aldosterone.

26. Which of the following conditions may be associated with a radiologically enlarged pituitary fossa?
1. Forbes-Albright syndrome of non-puerperal lactation;
2. primary hypothyroidism;
3. stenosis of the aqueduct of Sylvius;
4. internal carotid aneurysm;
5. Klinefelter's syndrome;
6. congenital adrenal hyperplasia;
7. craniopharyngioma.

Answers

1. 1, 2, 3, 6
2. 1, 2, 3, 4, 7
3. 3, 5
4. 1, 3, 4
5. 2, 3, 5
6. 2, 6
7. 1, 4, 5
8. 3, 4, 6, 7
9. 1, 2, 5, 6
10. 1, 2, 4, 7
11. 1, 3, 4, 5
12. 1, 2, 3, 4, 7
13. 2, 3, 4
14. 1, 2, 3, 4, 5, 9
15. 1, 2, 3, 6, 7
16. 1, 2, 5, 7
17. 1, 2, 3, 5
18. 3, 4, 5, 6, 7, 8
19. 2, 5
20. 2, 4, 5, 6
21. 1, 2, 4, 6, 7, 8
22. 1, 2, 4, 5, 6
23. 2, 4, 7, 8
24. 5
25. 5
26. 1, 2, 3, 4, 7

---

2. Posterior pituitary

HORMONES

1. Vasopressin (ADH) (arginine vasopressin, AVP, in all mammals except pig where it is lysine vasopressin).
2. Oxytocin.

Each is an octapeptide (MW ~ 1000), consists of pentapeptide ring and tripeptide amide side chain, and differs from the other in 2 amino acids.

Because of the close structural relationship there is some overlap of function between the 2 hormones. Oxytocin has very weak antidiuretic action; ADH is ~20% as effective as oxytocin on uterine muscle and milk discharge.

Active analogues of each have been synthesised, e.g. ornithine vasopressin, a potent vasoconstrictor with weak antidiuretic properties (relevant surgical use). DDAVP (1-desamino-8-D-arginyl vasopressin) is a synthetic ADH with potent antidiuretic properties and very weak vasoconstrictor action. Vasotocin has the oxytocin ring and the vasopressin side chain, and has functional activity of both hormones.

Formation and release (neurosecretion)
The hormones are synthesised in the endoplasmic reticulum of cell bodies of neurones of the paraventricular nucleus and the supraoptic nucleus, and are transported down the axoplasm of the hypothalamoneurohypophysial tracts to be stored in the terminals of these neurones in the three divisions of the neurohypophysis (especially the neural lobe, i.e. posterior pituitary) (see Fig. 2.1).

Oxytocin and vasopressin are produced, stored and released by separate neurones. Oxytocin-producing neurones are located almost entirely within the paraventricular nucleus, and vasopressin-producing neurones are chiefly confined to the supraoptic nucleus; however, both types of neurones are present in each nucleus. In transit down the axoplasm (axonal flow) each hormone is contained within large granular vesicles and complexed with a protein, neurophysin, the synthesis of which is closely associated with that of the hormones. There are 2 human neurophysins (MW ~ 10 000); oxytocin is complexed to neurophysin I, and vasopressin to neurophysin II.

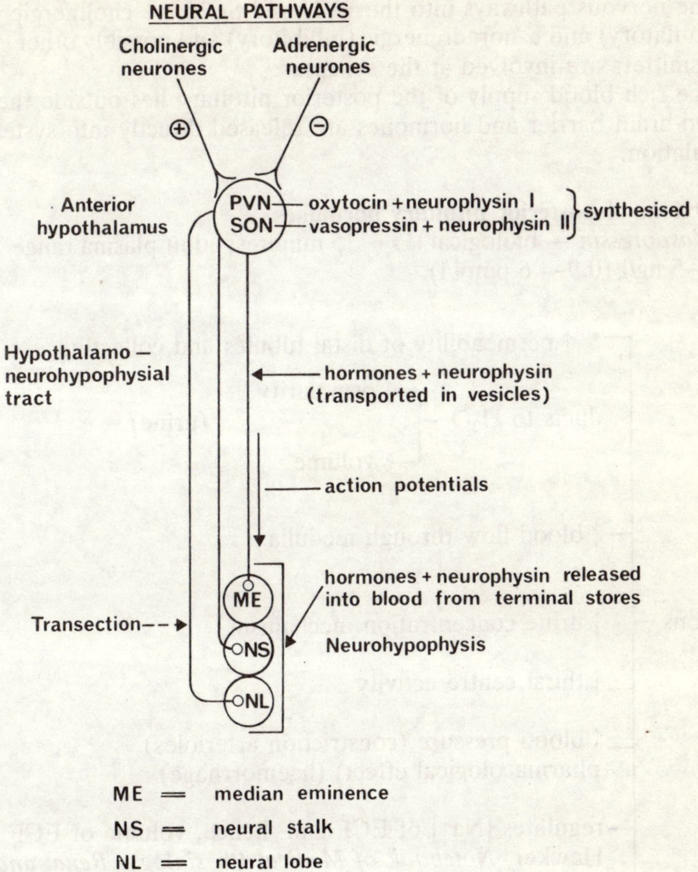

Fig. 2.1 Formation and release of oxytocin and vasopressin

Nerve impluses passing down the neurones regulate the release of hormones and neurophysins (separately) into the blood stream. Action potentials increase permeability of membrane to Ca^{++}, either directly or indirectly, through release of a chemical agent from vesicles aggregated in vicinity.

Transection (or injury) causes complete/partial diabetes insipidus according to level of lesion. Neurosecretory granules accumulate above lesion.

The hypothalamus contains about equal amounts of ADH and oxytocin. The evidence suggests that any stimulus which releases one hormone also releases the other, and releases about 200 times more oxytocin than ADH.

The nervous pathways into the PVN and SON are cholinergic (stimulatory) and α-noradrenergic (inhibitory) and possibly other transmitters are involved at the synapses.

The rich blood supply of the posterior pituitary lies outside the blood-brain barrier and hormones are released directly into systemic circulation.

Functions of posterior pituitary hormones

1. *Vasopressin* — biological t½ ~ 15 minutes; adult plasma range 1–5 ng/l (0.9–4.6 pmol/l).

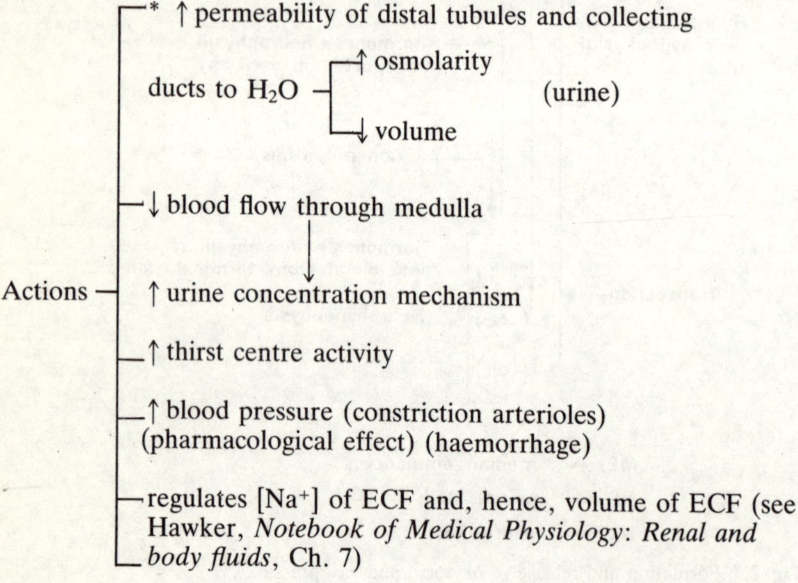

ADH acts on contraluminal surface (blood-side) of tubule, and its effect on cAMP generation (and water transport) is opposed by prostaglandin E.

ADH and water loss in urine of 100% H_2O filtered into proximal tubules ~ 20% enters distal convoluted tubules. In absence of ADH ~ 8% is reabsorbed (follows NaCl) in distal convoluted tubules and collecting ducts, and ~ 12% is lost in urine (e.g. $\dot{V}$ = 15 ml/min, and osmolarity of urine is 30 to 60 mmol/l; $\dot{V}$ = urine flow/min). With maximal ADH effect ~ 19% is reabsorbed in distal convoluted tubules and collecting ducts, and ~ 1% is lost in urine (e.g. $\dot{V}$ = 1 ml/min and osmolarity of urine = 1200 to 1400 mmol/1).

* Main physiological effect, distal tubules not affected in some species

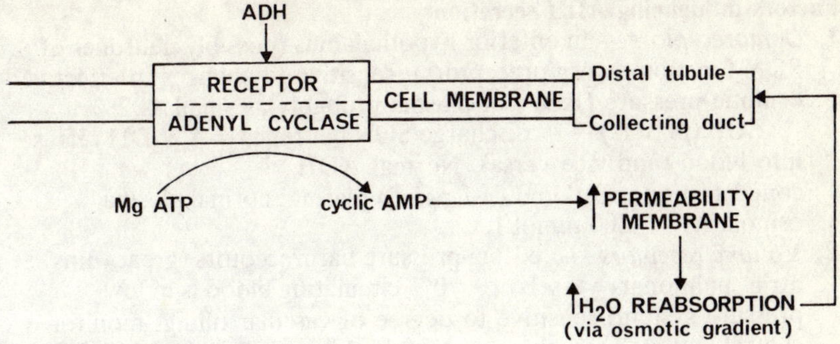

Fig. 2.2 Action of ADH
ADH → contraction cytoplasmic protein → ↑ pore size
↓
↑ permeability membrane

ADH influences (1) solute-free H_2O reabsorption (T_{CH_2O}) or alternatively (2) solute-free H_2O clearance (C_{H_2O}).

a. $T_{CH_2O} = C\ osm - \dot{V}$ (i.e. osmolar clearance – urine volume/min)
b. $C_{H_2O} = \dot{V} - C\ osm$

e.g. when C_{H_2O} is +ve (e.g. in H_2O diuresis), urine is hypotonic to plasma, and ADH effect is absent or minimal.

When C_{H_2O} is –ve (e.g. in dehydration), urine is hypertonic to plasma, and ADH effect is present and may be maximal.

C_{H_2O} is thus a measure of the gain or loss of H_2O by excretion of a concentrated or dilute urine, and it is mainly regulated by ADH (for further details see Hawker, *Notebook of Medical Physiology: Renal and body fluids*, Ch. 3).

2. *Oxytocin*

Actions —
- Contracts myoepithelial (smooth muscle) cells lining small ducts and alveoli → ↑ milk ejection into large ducts (sinuses) and nipple
- Contracts uterine muscle; oestrogens increase and progesterone decreases sensitivity of uterus to oxytocin. Oxytocin increases discharge and propagation of action potentials → ↑ uterine contractions. Uterus is especially sensitive to oxytocin during last month of pregnancy and during labour

Factors influencing ADH secretion

1. *Osmoreceptors* — in anterior hypothalamus (possibly neurones of SON function as osmoreceptors), sensitive to changes in effective osmotic pressure (EOP) of plasma probably as small as 2%. ↑ EOP (plasma) → ↑ discharge SON neurones → ↑ ADH release into blood (and vice versa). Normal ADH concentration ~ 3μU/ml (~1 pg/ml) plasma; normal plasma osmolarity ~ 300 mmol/1.
2. *Volume receptors* — i.e. low-pressure baroreceptors (great veins, atria, pulmonary vessels) (~ 70% circulating blood is in low-pressure system) sensitive to degree of vascular filling, monitor central venous pressure, and signal inhibitory impulses into SON. ↓ blood volume (↓ ECF) → ↓ stretch of receptors → ↓ vagal afferents → ↑ discharge SON neurones → ↑ ADH release into blood (and vice versa) (mechanism important in haemorrhage). Normal blood volume ~ 51 (ECF ~ 151). High-pressure baroreceptors (e.g. in carotid sinus) also respond if the hypovolaemia is sufficient to decrease rate of inhibitory impulses as the arterial blood pressure falls. Hypovolaemia is a more potent stimulus for ADH secretion than is plasma hyperosmolarity.

   Stimulatory effect of hypovolaemia on ADH secretion overrides inhibitory effects of hypo-osmolarity of plasma.
3. *Other factors* — emotional stimuli, stress, trauma, pain, surgery, nicotine, barbiturates, angiotensin, some anaesthetics, increased temperature, chlorpropamide and clofibrate increase ADH secretion.

   Alcohol, tranquilisers (e.g. reserpine, chlorpromazine), and some emotional states decrease secretion. These factors act by altering the balance between facilitatory (cholinergic) and inhibitory (adrenergic) influences.

Note: morphine and endogenous morphine-like peptides (endorphins and enkephalins) stimulate ADH release; morphine antagonists (e.g. oxilorphan) inhibit ADH release; hence, endogenous opiates may be involved in pain-induced ADH release.

Factors influencing oxytocin secretion

1. *Stimuli* — which release ADH also release oxytocin (but note, haemorrhage releases ADH mainly and little oxytocin).
2. *Suckling reflex* — suckling initiates integrated neuroendocrine reflexes mediated via the hypothalamus and pituitary which result in:
   a. oxytocin secretion → milk ejection and uterine contractions;
   b. prolactin secretion → milk secretion;
   c. ADH secretion → ↑ water reabsorption;
   d. stimulation of drinking centre → ↑ water intake;
   e. stimulation of feeding centre → ↑ food intake.

3. *Coital reflex* — oxytocin released during coitus is postulated to facilitate sperm transport into Fallopian tubes by increasing uterine contractions.
4. *Labour* — during 2nd stage labour afferent impulses from birth canal may stimulate PVN (and SON) to increase secretion of oxytocin which enhances uterine contractions. There is little evidence that oxytocin initiates onset of natural labour in women.

Factors influencing the thirst centre
These are shown in Figure 2.3.

The osmoreceptor mechanism for the release of ADH may be different from that for activation of the thirst mechanism, but the systems appear to be interlocked functionally.

There is a natural tendency for the body to become dehydrated from loss of fluids from various sources; hence there is a tendency

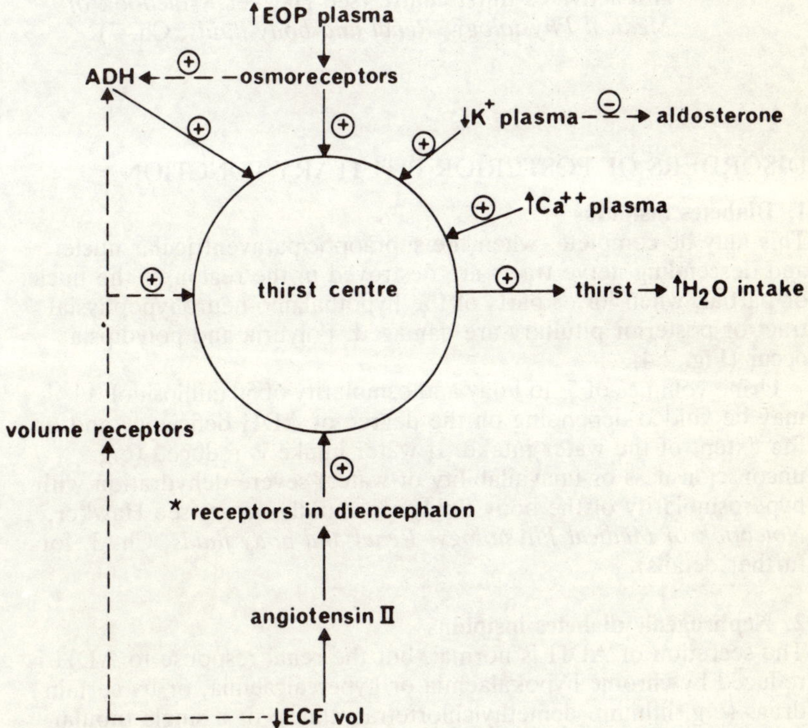

* = receptors are present in the subfornical organ and the organum vasculosum of the laminar terminalis, which are outside the blood-brain barrier

Fig. 2.3 Factors influencing the thirst centre (from Hawker (1982) *Notebook of Medical Physiology: Renal and body fluids*. Edinburgh: Churchill Livingstone)

for the osmolarity (Na concentration) of the ECF to increase. When the Na concentration rises sufficiently, the thirst mechanism is activated, and water is ingested: thus, the Na concentration is regulated.

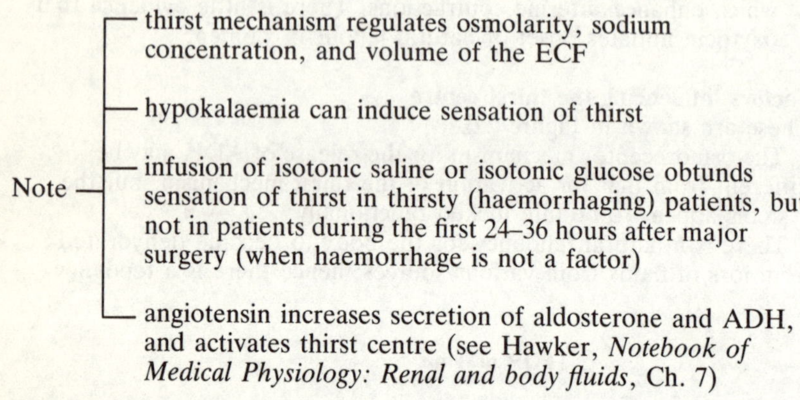

DISORDERS OF POSTERIOR PITUITARY FUNCTION

1. Diabetes insipidus

This may be complete, when the supraoptic/paraventricular nuclei and descending nerve tracts are destroyed in the region of the nuclei or partial, when lower parts of the hypothalamo-neurohypophysial tract or posterior pituitary are damaged. Polyuria and polydipsia occur (Fig. 2.4).

Urine volumes of 5–15 l/day and osmolarity of 50 milliosmoles/l may be voided depending on the degree of ADH deficiency and on the extent of the water intake. If water intake is reduced (e.g. unconsciousness or unavailability of water) severe dehydration with hyperosmolarity of the body fluids can rapidly occur (see Hawker, *Notebook of Medical Physiology: Renal and body fluids*, Ch. 3, for further details).

2. Nephrogenic diabetes insipidus

The secretion of ADH is normal, but the renal response to ADH is reduced by chronic hypokalaemia or hypercalcaemia, or by certain drugs (e.g. lithium, demethylchlortetracycline), or a single tubular genetic defect may be present.

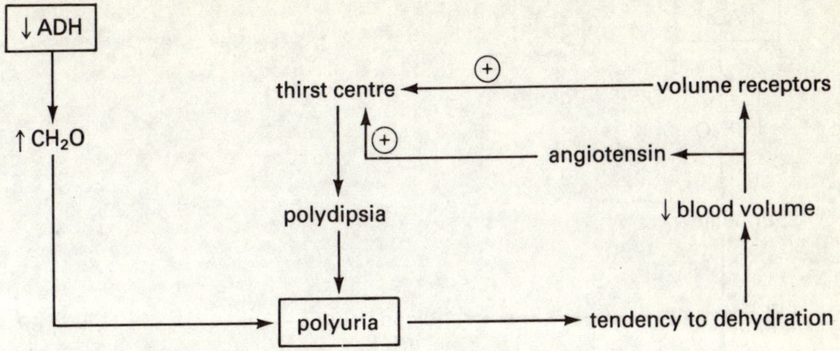

A vicious cycle is established:

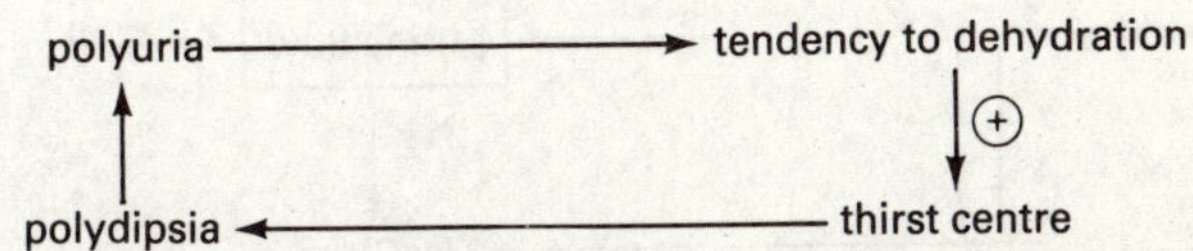

Fig. 2.4 Production of polyuria and polydipsia in diabetes insipidus

3. Syndrome of inappropriate ADH secretion (SIADH)
The secretion of ADH is excessive and inappropriate (i.e. the increased secretion is not in response to stimuli which normally increase secretion of ADH, e.g. hypovolaemia). ADH may be secreted by carcinoma of the bronchus or pancreas, or it may be secreted by the posterior pituitary, e.g. in patients with cerebral vascular disorders, or during intermittent positive pressure ventilation, or in response to certain drugs.

There is a tendency to onset of water intoxication (Fig. 2.5).

Note
- raised BP increases urine flow and overbalances reduced urine flow that is normally produced by ADH
- excessive loss of NaC1 in urine → hyponatraemia
- urine is highly concentrated (hypertonic to plasma) because of ↑ ADH

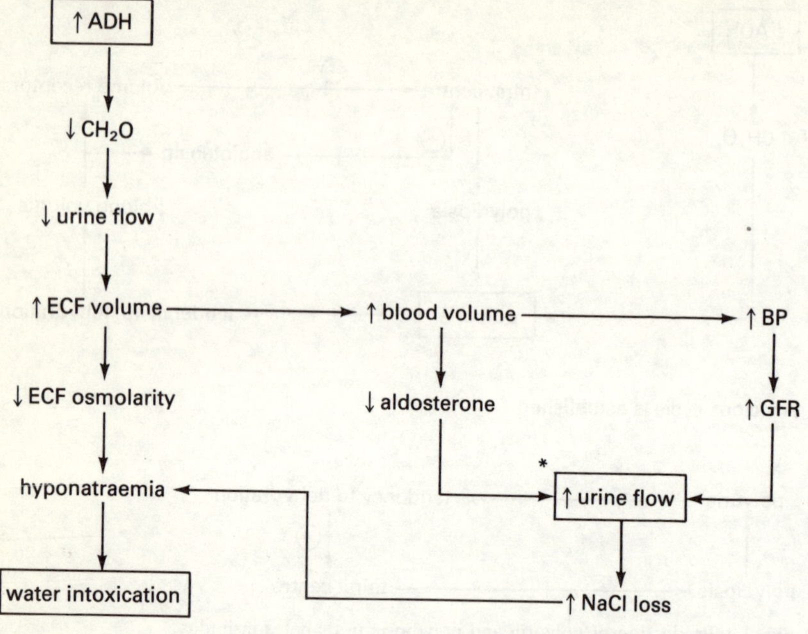

Fig. 2.5 Tendency to water intoxication in inappropriate ADH secretion

4. Combined lesions of the anterior and posterior pituitary
A lesion in the sella turcica which damages both the anterior and posterior pituitary is usually followed by a transient, slight polyuria, and not by diabetes insipidus as shown in Figure 2.6.

VASOPRESSIN FUNCTION TESTS (PRINCIPLES ONLY)

Water deprivation test
When normal subjects are dehydrated, vasopressin secretion is stimulated, the urine osmolarity is increased (e.g. 800 mmol/l), the plasma osmolarity changes little (e.g. from 285 to 295 mmol/l), and the C_{H2O} is –ve, i.e. urine is hypertonic to plasma.

In diabetes insipidus, the urine is hypotonic to plasma, i.e. C_{H_2O} is +ve, and plasma osmolarity is greater than 300 mmol/l (i.e. greater than that in normal subjects). The kidney is responsive to exogenous vasopressin.

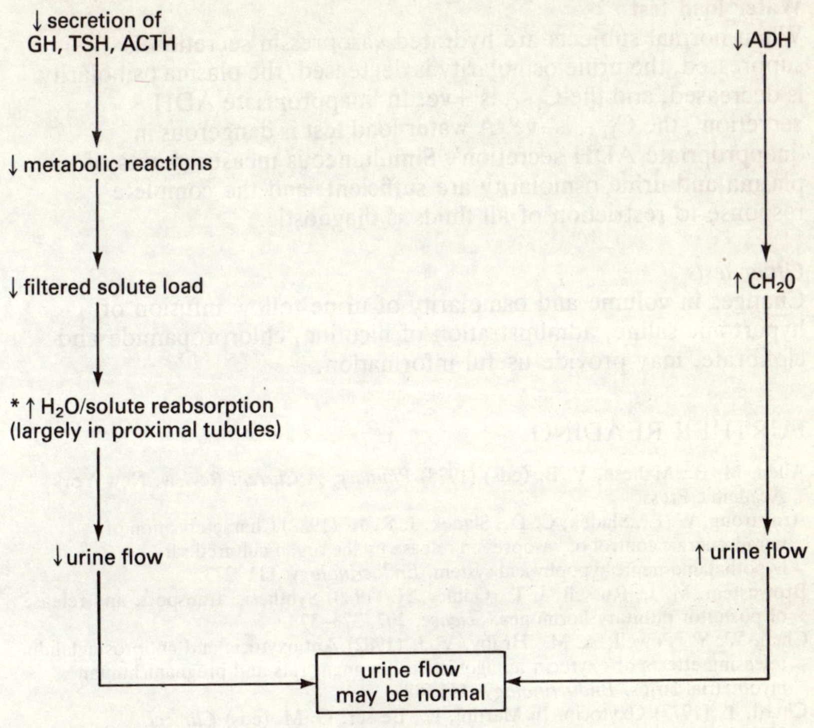

Fig. 2.6 Tendency to normal urine flow in combined lesions of anterior and posterior pituitary

Note: administration of GH or cortisol increases the tubular solute load and may induce frank diabetes insipidus.

When nephrogenic diabetes insipidus is present, the results conform with those for diabetes insipidus but the kidney is non-responsive to exogenous vasopressin.

With psychogenic polydipsia, at commencement of test the plasma osmolarity is normal or low (cf. high in diabetes insipidus). At the end of the test the C_{H_2O} is −ve, and response to exogenous vasopressin is much less than in diabetes insipidus.

Water load test

When normal subjects are hydrated vasopressin secretion is suppressed, the urine osmolarity is decreased, the plasma osmolarity is decreased, and the C_{H_2O} is +ve. In 'inappropriate ADH secretion', the C_{H_2O} is–ve. A water load test is dangerous in 'inappropriate ADH secretion'. Simultaneous measurements of plasma and urine osmolarity are sufficient, and the complete response to restriction of all fluids is diagnostic.

Other tests

Changes in volume and osmolarity of urine follow infusion of hypertonic saline, administration of nicotine, chlorpropamide and clofibrate, may provide useful information.

FURTHER READING

Allen, M. B., Mahesh, V. B. (eds) (1978) *Pituitary: A Current Review*. New York: Academic Press.

Armstrong, W. E., Sladek, C. D., Sladek, J. R. Jr. (1982) Characterisation of noradrenergic control of vasopressin release by the organ-cultured rat hypothalamo-neurohypophyseal system. *Endocrinology*, **111**, 273.

Brownstein, M. J., Russell, J. T., Gainer, H. (1980) Synthesis, transport, and release of posterior pituitary hormones. *Science*, **207**, 373–378.

Chan, W. Y., Powell, A. M., Hruby, V. J. (1982) Antioxytocic and antiprostaglandin-releasing effects of oxytocin antagonists in pregnant rats and pregnant human myometrial strips. *Endocrinology*, **111**, 48.

Chard, T. (1977) Oxytocin. In Martini, L., Besser, G. M. (eds) *Clinical Neuroendocrinology*, p. 569–583. New York: Academic Press.

Cobb, W. E., Spare, S., Reichlin, S. (1978) Neurogenic diabetes inspidius: management with DDAVP (1-Desamino-8-D arginine vasopressin). *Annals of Internal Medicine*, **88**, 183.

Defendini, R., Zimmerman, E. A. (1978) The magnocellular neurosecretory system of the mammalian hypothalamus. In Reichlin, S., Baldessarini, R. J. et al (eds) *The Hypothalamus*, vol. 56, p. 137–152. New York: Raven Press.

Dierickx, K., Vandesande, F. (1979) Immunocytochemical demonstration of separate vasopressin-neurophysin and oxytocin-neurophysin neurons in the human hypothalamus. *Cell Tissue Research*, **196**, 203.

Edwards, C. R. W. (1977) Vasopressin. In Martini, L., Besser, G. M. (eds) *Clinical Neuroendocrinology*, p. 527–567. New York: Academic Press.

Fitzsimons, J. T. (1972) Thirst. *Physiological Reviews*, **52**, 468–561.

Fitzsimons, J. T. (1979) The physiology of thirst and sodium appetite. *Monograph No. 35 of the Physiological Society*. Cambridge University Press.

Forsling, M. (1976) *Anti-diuretic Hormone*, vol. 1. Edinburgh: Churchill Livingstone.

Forsling, M. (1978) *Anti-diuretic Hormone*, vol. 2. Edinburgh: Churchill Livingstone.

Miller, M., Moses, A. M. (1977) Clinical states due to alteration of ADH release and action. In Phiebig, A. J. (ed) *Neurohypophysis* (International Conference, Key Biscayne, Florida, 1976). Basel: Karger.

Moore, G. J., Kwok, Y. C., Ko, E. M., Severson, D. L., Rosenior, J. C. (1982) Extended chain analogs of [Arginine[8]] Vasopressin as model prohormones: investigation of precursor-processing enzymes in extracts of the rat hypothalamus and neural lobe. *Endocrinology*, **111**, 1626.

Pickering B. T. (1978) Neurohyophysial hormones — comparative aspects. In Jeffcoate, S. L., Hutchinson, J. S. M. (eds) *The Endocrine Hypothalamus*, p. 213–227. London: Academic Press.

Robertson, G. L. (1977) The regulation of vasopressin function in health and disease. *Recent Progress in Hormone Research*, **33**, 333.

Robinson, A. G. (1977) Neurophysins. In Martini, L., Besser, G. M. (eds) *Clinical Neuroendocrinology*, p. 585–602. New York: Academic Press.

Walker, L. A., Whorton, A. R., Smigel, M., France, R., Frohlich, J. C. (1978) Antidiuretic hormone increases renal prostaglandin synthesis *in vivo*. *American Journal of Physiology*, **235**, F180.

Multiple choice questions

1. Vasopressin (ADH):
1. is produced largely in the supraoptic nuclei of the hypothalamus;
2. is not produced in the newborn until about 6 weeks after birth;
3. is transported in vesicles along the axoplasm of nerve fibres of the hypothalamo-hypophysial tract complexed to neurophysin; the same vesicles also contain oxytocin;
4. concentration increases in the plasma during acute haemorrhage (e.g. 1.5 l blood loss in 1 h): the magnitude of this response would be less if the vagus nerves to the heart were cut;
5. increases the permeability of cells of the distal convoluted tubules to water, thereby tending to render the urine hypertonic (to plasma) as it leaves these tubules;
6. in pharmacological doses has no effect on the myoepithelial cells in the breasts of the lactating mother;
7. action in cells of renal collecting ducts is mimicked by the addition of cyclic AMP to those cells;
8. is a protein with MW 20 000, approximately;
9. reaches its target cells in the nephron via the blood stream and not the tubular lumen fluid.

2. Oxytocin:
1. is essential in women for the onset and maintenance of natural labour;
2. is synthesised in the endoplasmic reticulum of cell bodies in the paraventricular nuclei of the hypothalamus;
3. is released from vesicles in nerve terminals in the neurohypophysis in response to action potentials conducted along the nerve fibres;
4. release is decreased by stimulation of cholinergic pathways onto the supraoptic nucleus;
5. induces milk ejection by contraction of myoepithelial cells which surround the small ducts and alveoli of the mammary glands;
6. secreted endogenously increases the discharge and propagation of action potentials in a progesterone-dominated uterus;
7. release is increased following a rise in the osmolarity of plasma from 300 to 320 mmol/l;
8. released in women during coitus probably facilitates sperm transport in uterus and Fallopian tubes.

3. Oxytocin:
1. acts in concert with testosterone and FSH for normal spermatogenesis;
2. induces milk secretion by a direct action on milk secreting cells;
3. appears to be released into the blood whenever ADH is released except during acute haemorrhage;
4. release is increased by pain and decreased by alcohol or water ingestion;
5. release can be increased in a mother as a conditioned response when her baby is regularly breast-fed;
6. is a polypeptide with a MW 1000, approximately;
7. action on the uterus is enhanced by oestrogen and decreased by progesterone;
8. structurally is similar to vasopressin.

4. With respect to neurohypophysial function:
1. the neurophysins are carrier proteins which are secreted from the posterior pituitary together with oxytocin and vasopressin;
2. high doses of oxytocin can cause antidiuresis;
3. oxytocin and vasopressin are synthesised in posterior pituitary cells;
4. the syndrome of inappropriate secretion of antidiuretic hormone is characterised by hyponatraemia, hypochloraemia and a normal or low serum creatinine level;
5. patients with diabetes insipidus can be treated successfully with chlorpropamide;
6. idiopathic diabetes insipidus is a disorder which usually has an abrupt onset;
7. hyponatraemia with clinical signs of dehydration is commonly the result of excessive vasopressin secretion.

5. ADH:
1. is synthesised in neurones different from those which synthesise oxytocin;
2. causes configurational changes of cytoplasmic proteins in cells in the distal tubules and collecting ducts and thereby increases membrane permeability to water at these sites;
3. like oestrogen, penetrates the cell membrane, binds with a cytoplasmic receptor which reacts with a nuclear protein receptor; this facilitates DNA transcription and m-RNA synthesis;
4. release is decreased when the arterial blood pressure rises, because impulses from baroreceptors in the carotid sinus inhibit its release;
5. release is decreased when the central venous pressure falls below normal;

6. constricts arterioles and, under physiological conditions, it thereby exerts an important effect on the peripheral resistance;
7. can constrict the vasa recta vessels in the renal medulla which would facilitate the urine concentration mechanism;
8. increases T_{CH_2O} (solute-free water reabsorption) in the kidney.

6. Consider the hormones produced in the hypothalamus:
1. they are synthesised in cell bodies of neurones in the hypothalamus;
2. all are released into the primary capillary plexus and are conveyed to the anterior pituitary via the hypothalamic-hypophysial portal circulation;
3. TRF increases release of TSH via a receptor-adenyl cyclase linked system;
4. oxytocin is stored in nerve terminals in the posterior pituitary and is released into the blood by oxytocin releasing factor which is a peptide;
5. they are small peptides;
6. they are complexed to neurophysin and are released into the blood in the hypothalamus;
7. GRF is essential for the release of GH;
8. CRF is subject to negative feedback regulation by ACTH.

7. In the syndrome of 'inappropriate ADH secretion':
1. the serum sodium level is high;
2. urine sodium excretion is increased;
3. remission can be expected with disappearance of the cause;
4. the level of urine sodium excretion is due to shutdown of aldosterone secretion and to diminished proximal tubular resorption;
5. the kidney is unusually sensitive to ADH.

8. With respect to ADH:
1. low-pressure baroreceptors (volume receptors) monitor the degree of vascular filling and respond to a fall in central venous pressure by initiating increased ADH secretion;
2. high-pressure baroreceptors (e.g. in carotid sinus) monitor changes in arterial blood pressure and respond to a fall in pressure by initiating increased ADH secretion;
3. during haemorrhage the secretion of ADH is increased; this is triggered by low-pressure and high-pressure baroreceptors;
4. in the presence of ADH, the filtrate which enters and leaves the distal convoluted tubules is isotonic with plasma;
5. ADH decreases the solute-free water clearance (C_{H_2O});
6. ADH increases K^+ exchange for Na^+ in distal tubules.

9. The secretion of ADH is inhibited by:
1. pain;
2. haemorrhage (acute);
3. whisky sufficient to cause mild intoxication;
4. ingestion of 1 litre of water within several minutes;
5. nicotine.

10. The following procedures are useful in the diagnosis of compulsive water drinking:
1. measurement of serum ADH;
2. observing response to water restriction over several days;
3. response to pitressin;
4. saline infusion test;
5. measurement of serum electrolytes;
6. measurement of urine electrolytes.

11. Oxytocin:
1. contracts myoepithelial cells in breasts;
2. aids milk delivery by dilating mammary ducts;
3. is complexed to neurophysin and enclosed in vesicles as it is transported down the hypothalamo-neurohypophysial tract;
4. is synthesised in the hypothalamus.

12. With respect to hormone synthesis: which is the odd one out?
1. adrenaline;
2. oxytocin;
3. GRF;
4. vasopressin;
5. aldosterone;
6. acetylcholine;
7. noradrenaline;
8. PIF.

13. The features of diabetes insipidus of neurohypophysial origin are:
1. tendency to dehydration;
2. tendency to hypovolaemia;
3. polydipsia;
4. polyphagia;
5. hypoosmolarity of ECF.

14. The features of inappropriate ADH secretion syndrome due to a bronchial carcinoma are:
1. urine hypertonic to plasma;
2. excessive expansion of ECF volume and commonly associated with oedema;
3. mild expansion of ECF volume;
4. increased blood pressure;

5. hyponatraemia;
6. reduced urine volume;
7. tendency to water intoxication.

15. Nephrogenic diabetes insipidus may be associated with:
1. normal secretion of ADH;
2. decreased secretion of ADH;
3. hypokalaemia;
4. single tubular genetic defect;
5. impairment of countercurrent mechanism for concentrating the urine;
6. renal tubules which are unresponsive to ADH.

Answers

1. 1, 4, 7, 9
2. 2, 3, 5, 7, 8
3. 3, 4, 5, 6, 7, 8
4. 1, 2, 4, 5, 6
5. 1, 2, 4, 7, 8
6. 1, 3, 5, 8
7. 2, 3, 4
8. 1, 2, 3, 5
9. 3, 4
10. 1, 2, 3, 4
11. 1, 3, 4
12. 5
13. 1, 2, 3
14. 1, 3, 4, 5, 7
15. 1, 3, 4, 5, 6

---

3. Thyroid

THYROID HORMONE BIOSYNTHESIS

See Figures 3.1 and 3.2.

1. Inorganic iodide (I^-) is removed from plasma by active transport (thyroid to serum concentration ratio, T/S = 25 or more depending on cell activity), is concentrated in follicular lumen, and reactions then proceed at the cell-colloid interface.

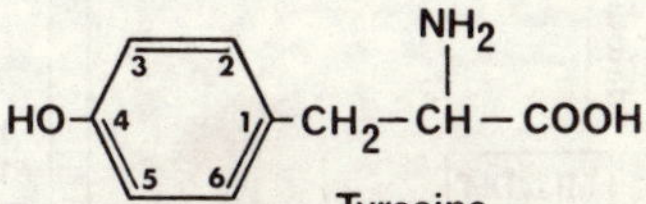

Tyrosine

3 – monoiodotyrosine (MIT)

3,5 – diiodotyrosine (DIT)

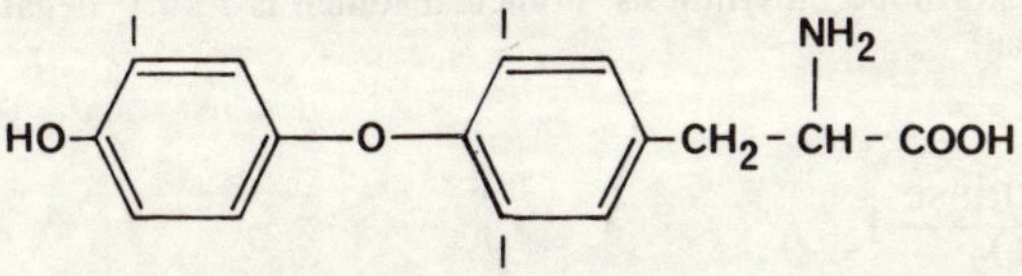

3,5,3′ – triiodothyronine (T_3)

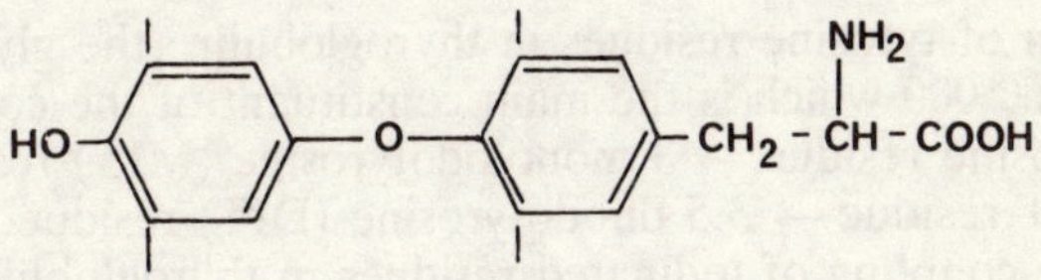

3,5,3′,5′ – tetraiodothyronine (thyroxine, T_4)

Fig. 3.1 Structural formulae of thyroid hormones and precursors

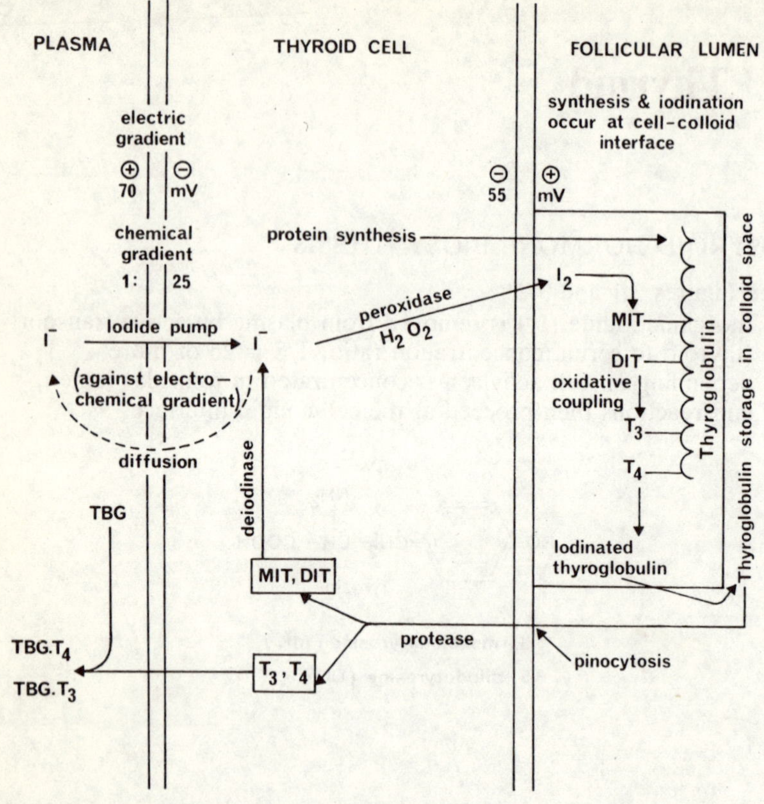

Fig.3.2 Thyroid hormone biosynthesis. Follicular lumen is 15 mV negative with respect to plasma

2. $2\,I^- \xrightarrow[H_2O_2]{\text{peroxidase}} I_2$

   An activation step occurs prior to the iodination of tyrosine, although the precise chemical state of the activated iodine is uncertain.
3. Iodination of tyrosine residues in thyroglobulin (the glycoprotein of MW 660 000 which is the main constituent of the colloid).
   $\frac{1}{2}I_2$ + tyrosine residue → 3 monoiodotyrosine (MIT) residue
   $\frac{1}{2}I_2$ + MIT residue → 3,5 diiodotyrosine (DIT) residue.
4. Oxidative coupling of iodinated residues in thyroglobulin. This is a slow, nonenzymatic reaction which proceeds during storage of colloid.
   MIT residue + DIT residue → 3,5,3′ triiodothyronine (T3) residue + alanine residue
   DIT residue + DIT residue → 3,5,3′,5′ tetraiodothyronine (thyroxine, T4) residue + alanine residue.

5. Pinocytosis and proteolytic degradation of thyroglobulin is necessary before hormone stored in the colloid can be released. Lysosomes containing proteases coalesce with pinocytic vacuoles; the T3 and T4 released diffuse into plasma, while MIT and DIT are deiodinated releasing iodide, most of which is retained by the gland to be reincorporated into thyroglobulin.
6. Most of the circulating T3 does not originate in the thyroid, but from deiodination of T4 in liver and muscle. T3 is the principal active thyroid hormone in postnatal life.

THYROID STIMULATORS

See Figure 3.3.

Thyroid-stimulating hormone (TSH) and thyroid-stimulating auto-antibodies (TSAb) (synonym: thyroid-stimulating immunoglobulins, TSI) act via adenyl cyclase (see p. 7) to stimulate iodide trapping. hormone synthesis and release, intermediary metabolism and cell growth. The depletion of colloid seen in the stimulated gland is due to enhanced degradation which exceeds the rate of synthesis.

Thyroid-stimulating auto-antibodies are found in patients with Graves' disease. The first of this class of antibodies to be discovered was the 'long-acting thyroid stimulator' (LATS) so named because of the time-course of response in a mouse bioassay. The immunological cross-reaction between human TSAb and mouse thyroid is variable; hence the correlation between LATS assay and thyroid function is poor. The discovery of thyroid-stimulating antibodies which are undetectable in the mouse assay but which react specifically with human thyroid tissue has resolved previous difficulties in accounting for hyperthyroidism in Graves' disease. These stimulators, known as LATS-protector (name derived from the competitive binding assay method which detects it) are present in all thyrotoxics and correlate well with thyroid function. The introduction of radioreceptor assay methods has led to the former terminology being replaced by the unitary concept TSAb.

Patho-physiology of thyrotoxicosis

Graves' disease shows some genetic predisposition, with histocompatibility antigens HLA B8 and DR3 associated with increased risk, and male gender having reduced risk. The thyroid in Graves' disease is believed to be normal apart from the effects of circulating TSAb and infiltrating lymphocytes. After subtotal thyroidectomy or ^{131}I treatment the thyroid remnant can assume a normal level of function under control of the pituitary-thyroid servo-system. There is a tendency for the auto-immune process to remit spontaneously, accounting for the long term success of medical treatment in a certain percentage of cases and for the slow auto-immune destruction of the gland leading eventually to myxoedema.

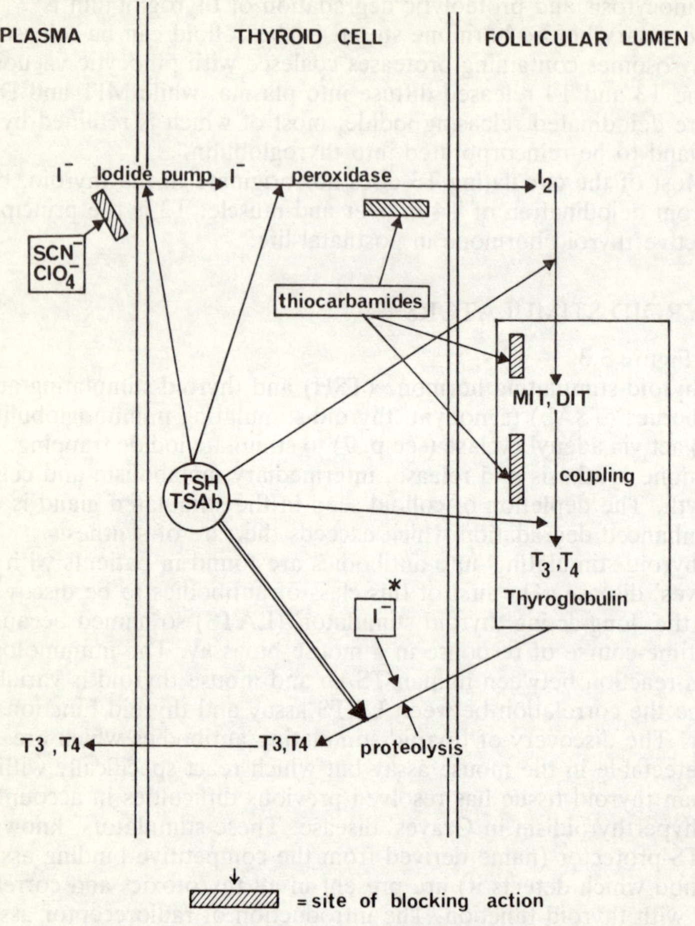

Fig.3.3 Actions of TSH, TSAb, and thyroid blocking agents

The ocular and dermatological manifestations of the disease are thought to be auto-immune in origin though there is no simple relationship between these signs and the presence of thyroid-stimulating antibodies. The pituitary is no longer believed to be involved in the ophthalmopathy or in the maintenance of hyperthyroidism.

THYROID BLOCKING AGENTS (ANTITHYROID SUBSTANCES, GOITROGENS)

See Figure 3.3.

Iodide pump inhibitors (Pseudohalides)
Thiocyanate (SCN^-) and perchlorate (ClO_4^-) competitively inhibit I^- uptake.

Inhibitors of organification
Thiocarbamides (e.g. carbimazole). Block $I^- \rightarrow I_2$, iodination of tyrosine to produce MIT and DIT, coupling of iodotyrosines to form T3, T4.

Iodide transport is unaffected by these drugs, but iodide diffuses back into plasma because of failure of organification.

Both pseudohalides and thiocarbamides →
- ↓ total thyroidal iodine
- ↓ ^{131}I uptake value*

But T/S for iodide (I^-) is
- ↓ by pseudohalides
- ↑ by thiocarbamides, since iodide accumulates rather than being promptly organified.

Iodide
In thyrotoxicosis high doses temporarily (7 to 14 days) inhibit the biosynthesis of thyroid hormones, thyroglobulin degradation and hormone release, and result in a decrease in vascularity and thyroid size with an increase in colloid storage. This effect is not seen in euthyroid subjects. Iodine goitre (cough mixtures, kelp) is a rarity associated with a genetically determined enzyme defect.

Goitre production
Iodine deficiency
Antithyroid drugs
Thyroidal enzyme defects
→ ↓ T3, T4 synthesis → ↑ TRF → ↑ plasma TSH → goitre

* Uptake = iodide and iodine in gland in both organic and inorganic forms (in absence of antithyroid drugs, radio-iodine uptake comprises 98% organic iodine and 2% I^-)

Goitre may result from genetically induced deficiencies in any of the enzymes in the biosynthetic pathway (including deiodinase) leading to inadequate thyroid hormone production. The occurrence of goitre in deiodinase deficiency emphasises the importance of this enzyme's role in normal iodine conservation.

Metabolism of 70 μg iodine/day (equivalent to synthesis and degradation of 100 μg T4) (see Fig. 3.4).

Minimum intake to replace obligatory losses and maintain euthyroidism is 20 μg iodine/day.

Approximately 5 mg I_2 in thyroid (90% of body total)

Approximately 98% I_2 in gland is organically bound — 35% T4, 5% T3, 34% DIT, 24% MIT

Approximately 2% I_2 in gland is in inorganic form

Secreted hormones — > 95% T4 (80–100 μg/day); < 5% T3 (5 μg/day)

T4 -t½ in blood, 6–7 days
(3 to 4 days in hyperthyroidism, 9 to 12 in myxoedema)
T3 -t½ in blood, 2 days

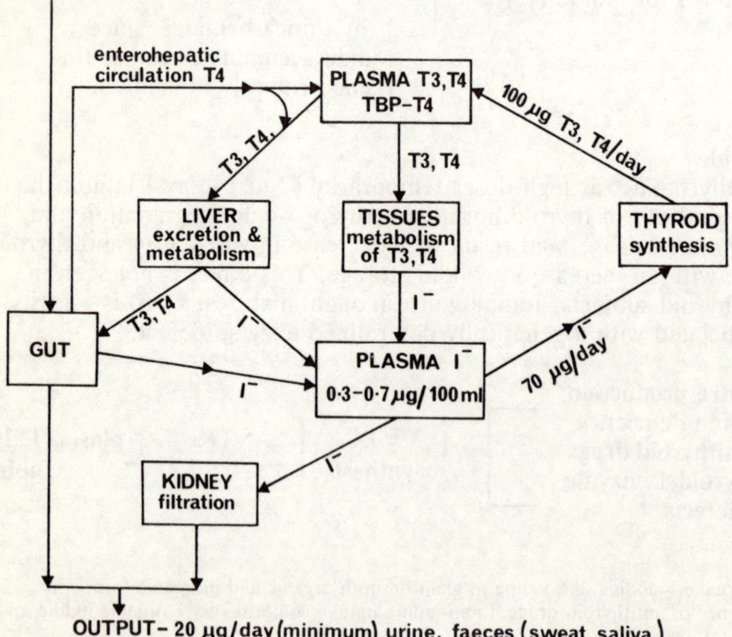

Fig. 3.4 Iodine metabolism

Key: TBP-T4 = T4 bound to thyroxine binding proteins

Metabolism of T4 and T3

Both T4 and T3 are deaminated (especially in kidney and liver) and deiodinated (especially in liver and muscle) with some conversion of T4 to T3 and recirculation of liberated iodide to the thyroid. A competing pathway converting T4 to the metabolically inactive reverse-T3 may provide some regulation of thyroid hormone utilisation at tissue level. Thiocarbamides inhibit the deiodination of T4 in peripheral tissues.

T4 and T3 are conjugated with sulphates and glucuronides (liver) and excreted in bile; some enterohepatic circulation of free hormones occurs following hydrolysis of the conjugated forms in the small intestine.

ACTIONS OF THYROID HORMONES

1. ↑ activity of Na^+ pump ATP-ase (ATP → ADP + energy) of cell membranes in all adult cells except brain and retina and in all cells in infants and neonates → ↑ metabolic rate.
2. ↑ consumption (breakdown) of ATP is met by:
   ↑ transport of ADP into mitochondria → ↑ synthesis of ATP
   ↑ number and size of mitochondria and activity of oxidative enzymes → ↑ synthesis of ATP
   ↑ m-RNA-ribosome activity → ↑ synthesis of metabolic enzymes.
3. ↑ consumption of ATP leads to ↑ cell utilisation of:
   O_2 (↑ HbO_2 dissociation, favoured by ↑ 2,3 DPG concentration); glucose (oxidation → ATP; oxidative phosphorylation (see p. 138) and increasing amounts of fat and protein if glucose supplies become inadequate;
   vitamins (B group, B12, C, A).
4. Overall increase in turnover of substrates and metabolic intermediates:
   ↑ calorific intake and/or weight loss
   ↑ glucose turnover (glycogenolysis, gluconeogenesis)
   ↑ lipid turnover (plasma cholesterol decreases due to increased removal by the liver despite increased hepatic synthesis)
   ↑ protein catabolism
   ↑ conversion of carotene to vitamin A by the liver, hence the carotenaemia seen in myxoedema.
5. Permissive effects:
   required for adequate protein synthesis in growing and differentiating tissues, potentiate effects of GH and insulin, potentiate catecholamine effects — thermogenesis (fat cells and muscle), CVS responses (tachycardia, vasodilatation/vasoconstriction), CNS (reticular activating system).

6. Effects in thyrotoxicosis and with excessive dosage of administered T4

Negative balance with respect to:
Calories
Fat

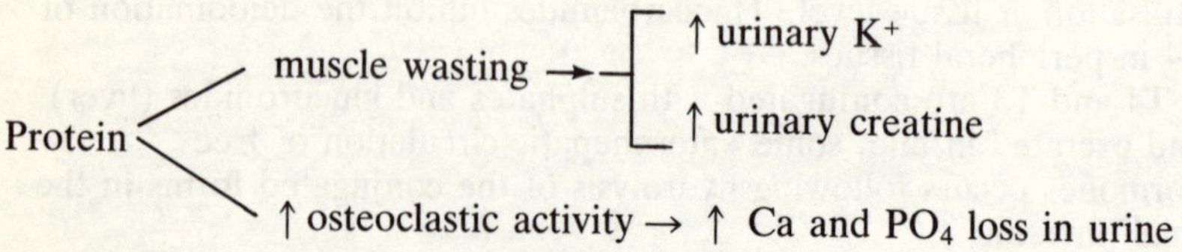

CONTROL OF THYROID SECRETION

See Figure 3.5.

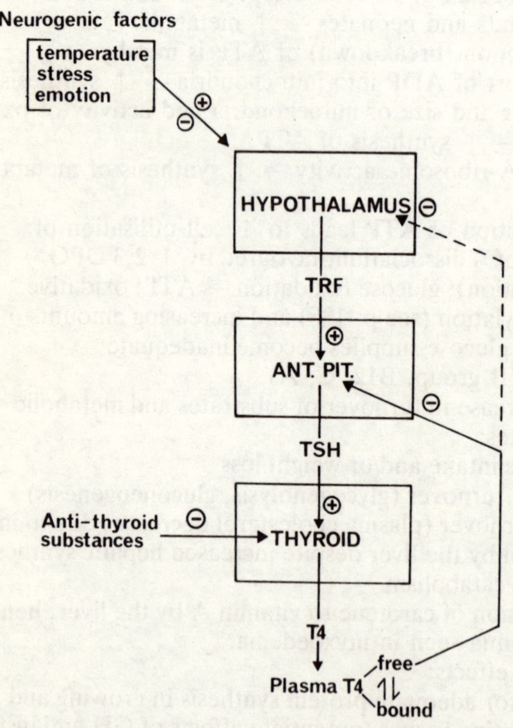

Fig. 3.5 Regulation of thyroid secretion

THYROID HORMONE TRANSPORT IN BLOOD

T4 is strongly bound to specific plasma proteins (< 0.1% is free). Adult range 4.3 to 11.2 μg/100 ml (55 to 150 nmol/l) (total).

T3 is less strongly bound (< 0.5% is free). Adult range 0.6 to 1.8 ng/ml (0.9 to 2.8 nmol/l) (total).

It is the free (non-protein bound) T4 (10 to 30 pmol/l) and T3 which are metabolically active and which mediate the negative feedback control of TSH secretion. Consequently alterations in binding capacity of plasma proteins may be associated with parallel changes in total plasma T4 without altering free T4 levels or the thyroid status of the individual.

The level of thyroxine binding protein (TBP) increases during pregnancy, oestrogen administration and while taking some oral contraceptives.

↑ TBP → ↑ total T4 but normal free T4.

Hormones are bound to:

Thyroxine binding globulin (TBG). Most important carrier of T4 (and T3) ~ 90% T4 is bound to TBG. Binding capacity 22 μg/100 ml (~ 33% saturated). TBG has a greater affinity for T4 than for T3. Approximately 80% of T3 is bound to TBG. Binding can be decreased by various drugs (salicylates, phenytoin and bishydroxycoumarin). TBG values 12 to 30 mg/l.

Thyroxine binding pre-albumin (TBPA) binds some T4 (~ 10%).

Albumin. T4 is bound only when the capacity of the former binding proteins is exceeded. Approximately 20% of T3 is bound to albumin.

TBG deficiency may occur and is apparently due to deficiency of a single gene locus on the X chromosome. Deficiency of TBG is associated with euthyroidism, decreased total plasma T4 level, and increased T3-resin uptake.

THYROID FUNCTION TESTS

Total thyroxine estimation

Protein bound iodine (PBI) and butanol extractable iodine (BEI) measure iodine content of organic compounds in plasma (~ 95% T4, remainder T3, traces of MIT and DIT).

PBI = 4–8 μg/100 ml (corresponding approximately with 6 to 12 μg T4/100 ml since T4 is 2/3 iodine). Values raised after high iodine intake (radiological contrast media; iodide-containing cough mixtures).

These iodometric methods have been superseded by determination of total thyroxine (free and protein bound) by competitive binding radioassay (similar to radioimmunoassay, but using TBG as the binding agent in place of antibody).

Values of total thyroxine (normal range 4.3 to 11.2 μg/100 ml or 55 to 150 nmol/l) are increased in thyrotoxicosis and depressed in myxoedema, but are also increased in euthyroid states associated with increases in TBG, e.g. pregnancy, oestrogen administration.

T3-resin uptake
An index of the degree of saturation of TBG. Normal range 32 to 45%. Increased in thyrotoxicosis, decreased in myxoedema and in conditions (oestrogen therapy, pregnancy) where TBG concentrations are increased.

Free thyroxine estimate
These techniques provide an index of metabolically available thyroxine by combining indices of total thyroxine concentration and protein binding capacity. Normal range 90 to 110% (absolute range is 10 to 30 pmol/l). Increased in thyrotoxicosis, decreased in myxoedema, unaffected by oestrogen therapy and pregnancy.

Radioimmunoassay of TSH
Elevation of plasma TSH is a particularly sensitive index of hypothyroidism. Normal range is < 1 to 5 mIU/l.

TRF test
The injection of a dose of TRF into normal subjects causes a prompt rise in plasma TSH, whereas the chronically suppressed thyrotroph cells of the thyrotoxic patient fail to respond to the standard dose. The test is therefore useful in confirmation of thyrotoxicosis and also as a general test of pituitary reserve.

Radioimmunoassay of T3
is useful in confirming a diagnosis of T3 thyrotoxicosis, where the thyroid in Graves' disease secretes excessive T3 without much increase in T4 output (uncommon).

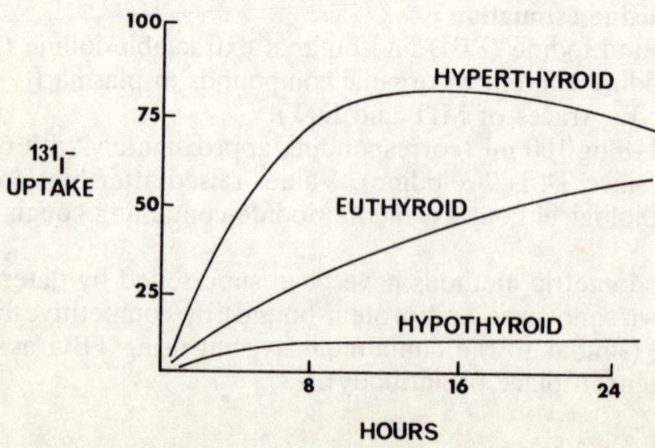

Fig. 3.6 Time-course of thyroid radioiodine uptake

Radioiodine uptake
^{131}I ($t\frac{1}{2}$ 8 days) is most commonly used, in oral dosages of ~ 10μCi (only 1/100 or less of the therapeutic dose used to produce thyroid destruction). ^{131}I produces γ rays which pass readily through tissues and can be detected externally as well as β rays which have a very short range in tissues and produce a local destructive effect. Normal range 15 to 45% of the dose in 24 hours. In the hyperthyroid state the initial uptake rate and the peak uptake are elevated. The level of iodide in the diet and renal function will influence radioiodine uptake ↑ dietary iodine → depressed uptake) (see Fig. 3.6).

Radioiodine uptake reflects the activity of the iodide pump and organification steps in hormone biosynthesis and will be depressed by antithyroid drugs.

Defects in other phases of hormone biosynthesis may destroy the relationship between thyroid status and iodine uptake, e.g. goitrous hypothyroidism due to deiodinase deficiency and severe dietary iodine deficiency with elevated rather than reduced uptake. The appearance of radioiodine in plasma thyroxine 24 hours after the test dose can be employed as a test of the overall functional status of the gland.

DISORDERS OF THYROID FUNCTION

Hypothyroidism

1. Primary hypothyroidism

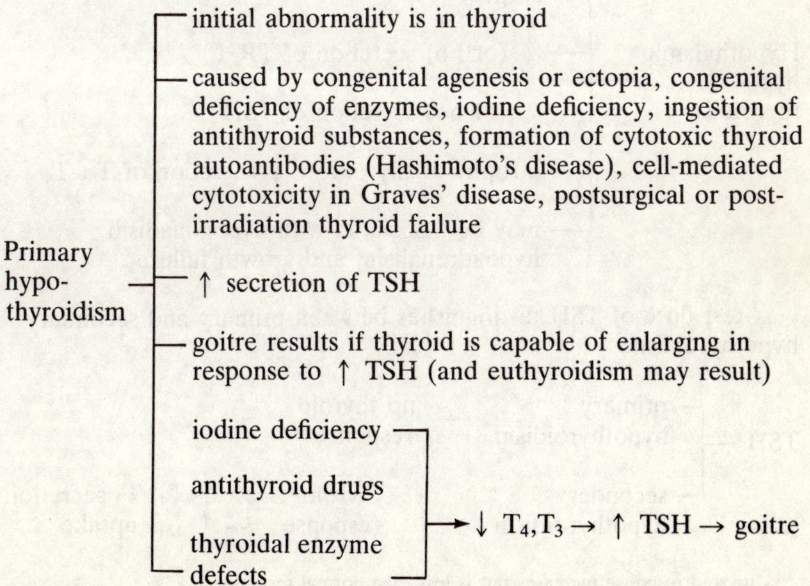

2. Secondary hypothyroidism

a. pituitary hypothyroidism

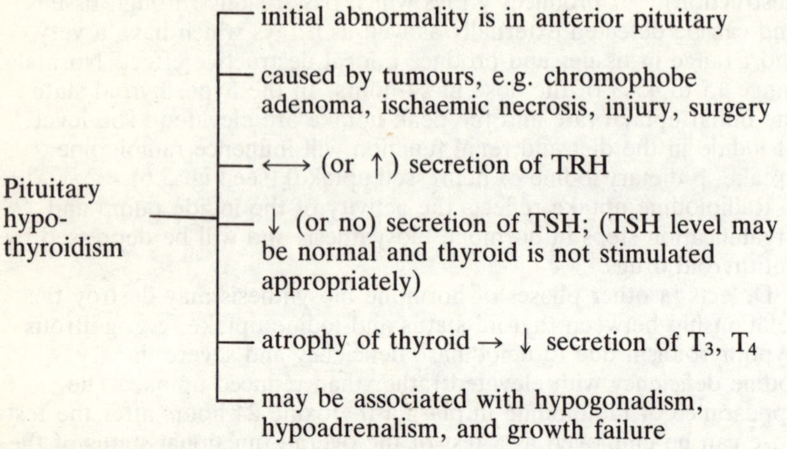

b. hypothalamic hypothyroidism

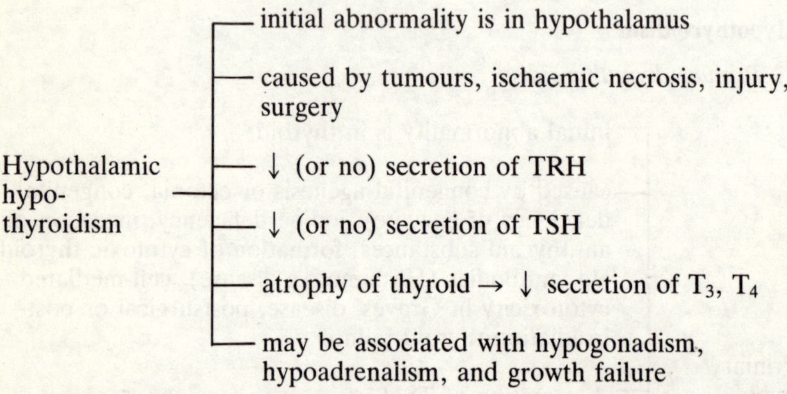

A test dose of TSH distinguishes between primary and secondary hypothyroidism:

TSH —
- primary hypothyroidism → no thyroid response
- secondary hypothyroidism → ↑* thyroid response —
  - ↑ T_4, T_3 secretion
  - ↑ $_{131}$I uptake

* = thyroid response increases but is less than normal response

Physiological effects of hypothyroidism

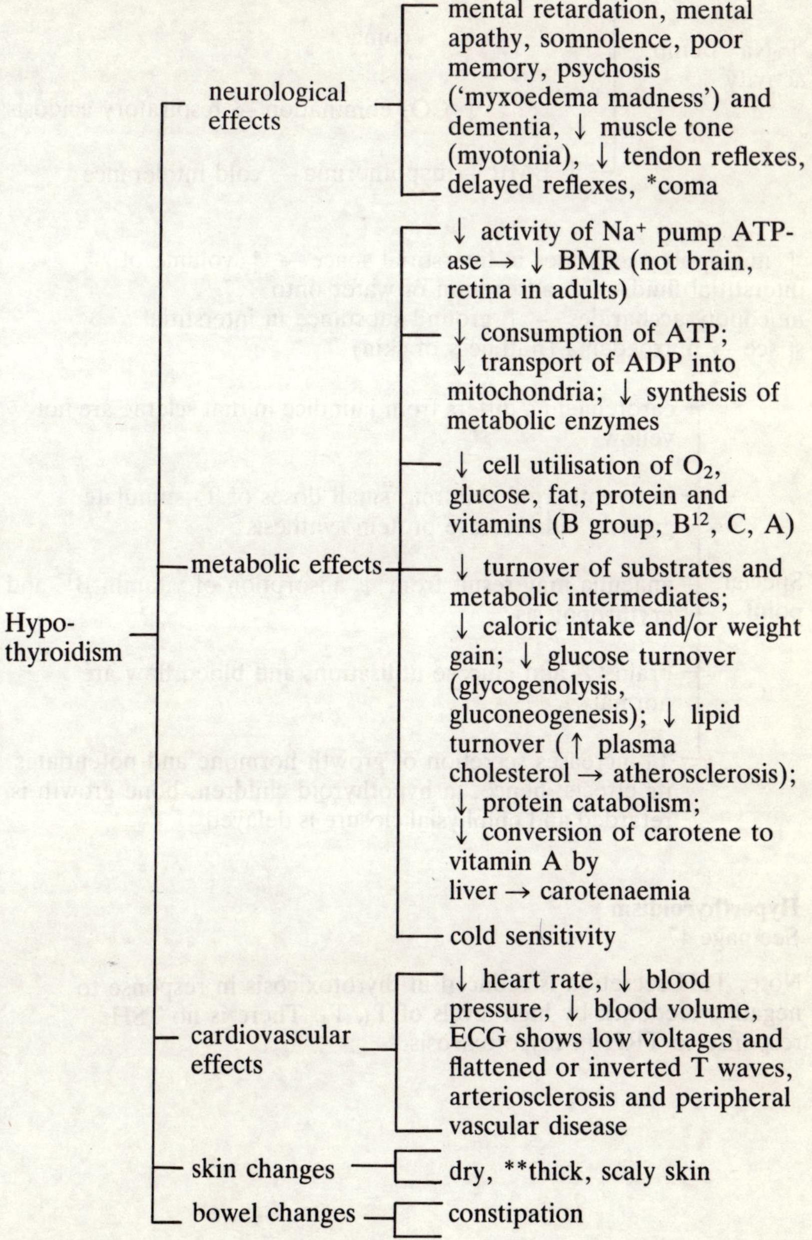

* = coma and hypothermia may follow
** = myxoedema is a nonpitting thickening of subcutaneous tissues

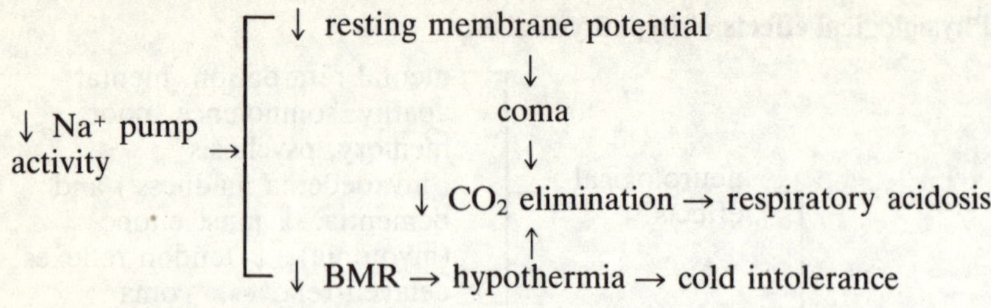

↑ mucopolysaccharides in interstitial space → ↑ volume of interstitial fluid → ↑ adsorption of water onto mucopolysaccharides → ↑ ground substance in interstitial space → myxoedema (puffiness of skin)

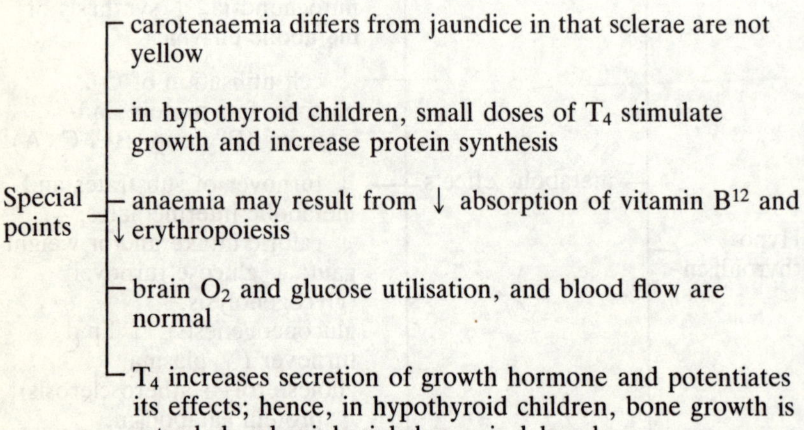

Hyperthyroidism

See page 47

Note: TSH secretion is reduced in thyrotoxicosis in response to negative feedback by high levels of T_3, T_4. There is no TSH response to TRH in thyrotoxicosis.

Physiological effects of hyperthyroidism

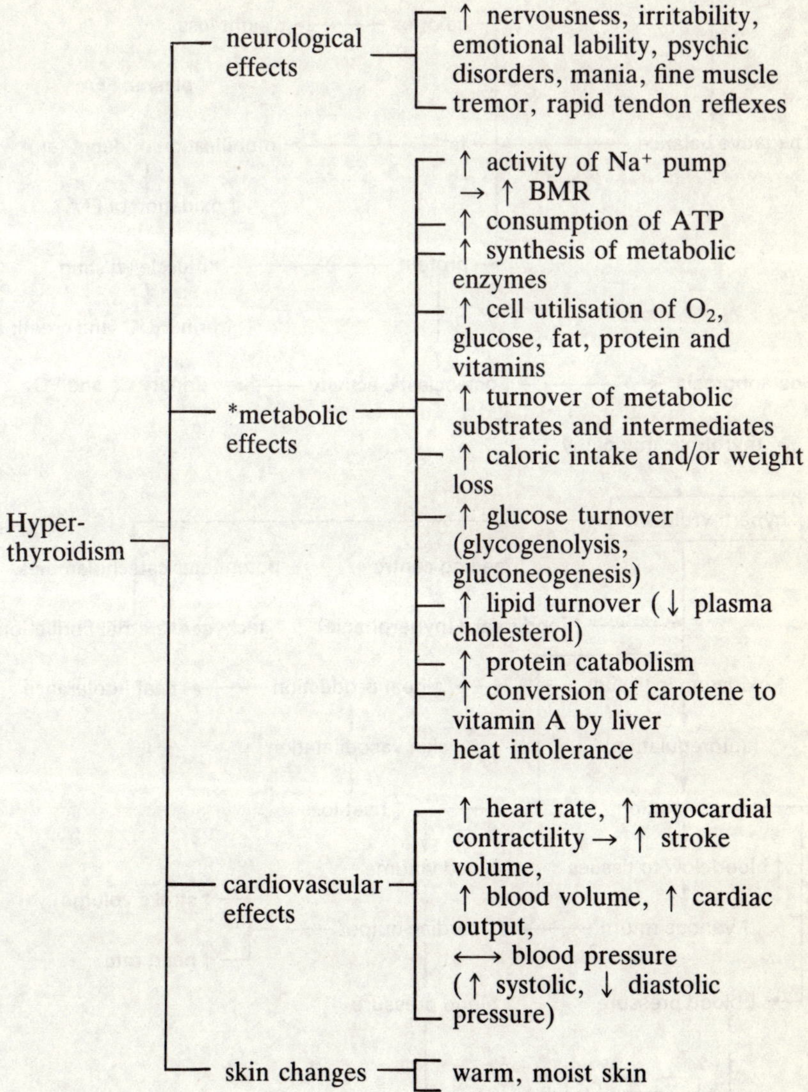

* = converse of effects present in hypothyroidism and reflect increased activity of T_3, T_4

hyperthyroidism
negative balance
calories
weight loss
↑ plasma FFA
fat
mobilisation of depot fat
↑ oxidation of FFA
protein
*muscle wasting
↑ urinary K+ and creatine
osteoporosis
↑ osteoclastic activity
↑ urinary Ca and PO4

* = thyrotoxic myopathy

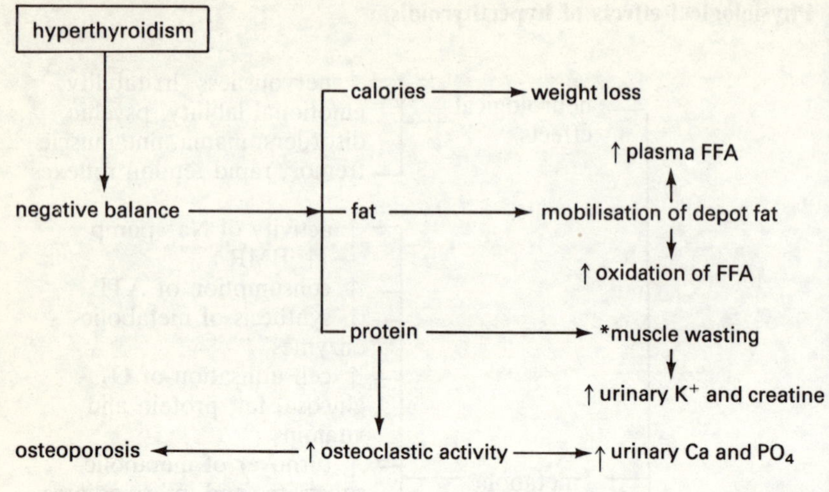

hyperthyroidism
+
feeding centre
potentiates catecholamines
↑ food intake (hyperphagia)
tachycardia/atrial fibrillation
↑ oxidation in tissues
↑ heat production
heat intolerance
(autoregulation)
skin vasodilatation
vasodilatation
↑ heat loss
↑ blood flow to tissues
↑ blood volume
↑ stroke volume
↑ venous return
↑ *cardiac output
↑ heart rate
↓ blood pressure
↑ blood pressure
↔ mean blood pressure
↑ pulse pressure

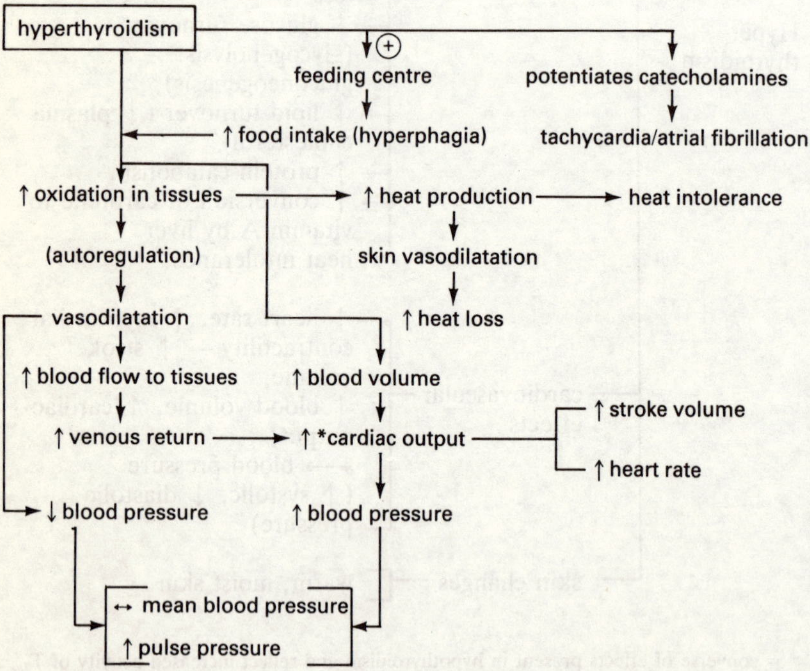

* = high output heart failure may develop

Fig. 3.7 Cardiovascular effects in hyperthyroidism

FURTHER READING

Bastomsky, C. H. (1974) Thyroid iodide transport. In Greer, M. A., Soloman, D. H. (eds) *Handbook of Physiology*, section 7: Endocrinology vol. III, Thyroid. Baltimore: Williams & Wilkins.

Bayer, M. F., McDougall, I. R. (1980) Radioimmunoassay of free thyroxine in serum: comparison with chemical findings and results of conventional thyroid-function tests. *Clinical Chemistry*, **26**, 1186.

Becker, D. V. (1978) Tests of peripheral thyroid hormone action: metabolic indices. In Werner, S. C., Ingbar, S. H. (eds) *The Thyroid*. Hagerstown: Harper & Row.

Braverman, L. E. (1978) Normal and abnormal responses to iodine: disorders of iodine excess. In Werner, S. C., Ingbar, S. H. (eds) *The Thyroid*. Hagerstown: Harper & Row.

Burman, K. D. (1978) Recent developments in thyroid hormone metabolism: interpretation and significance of measurements of reverse T_3, 3,3′-T_2, and thyroglobulin. *Metabolism*, **27**, 615.

D'Agostino, J., Henning, S. J. (1982) Postnatal development of corticosteroid-binding globulin: effects of thyroxine. *Endocrinology*, **111**, 1476.

DeGroot, L. J. (1979) Mechanism of action of thyroid hormone. In Ekins, R., Faglia, G., Pennisi, F., Pinchera, A. (eds) *Free Thyroid Hormones*. Amsterdam: Excerpta Medica.

Dunn, A. D., Dunn J. T. (1982) Thyroglobulin degradation by thyroidal proteases: action of purified cathepsin D. *Endocrinology*. **111**, 280.

Field, J. B. (1978) Pituitary thyrotropin: mechanism of action. In Werner, S. C., Ingbar, S. H. (eds) *The Thyroid*. Hagerstown: Harper & Row.

Green, W. L. (1978) Mechanisms of action of antithyroid compounds. In Werner, S. C., Ingbar, S. H. (eds) *The Thyroid*. Hagerstown: Harper & Row.

Ingbar, S. H., Borges, M. (1979) Peripheral metabolism of the thyroid hormones. In Ekins, R., Faglia, G., Pennisi, F., Pinchera, A. (eds) *Free Thyroid Hormones*. Amsterdam: Excerpta Medica.

McKenzie, J M., Zakarija, M. (1977) LATS in Graves' disease. *Recent Progress in Hormone Research*, **33**, 29.

Oppenheimer, J. H. (1979) Thyroid hormone action at the cellular level. *Science*, **203**, 971.

Schwartz, T. B. (ed.) (1975) The thyroid gland — hyperthyroidism. In *Year Book of Endocrinology*, p. 157–173. Chicago: Year Book Medical Publishers.

Segal, J., Ingbar, S. H. (1980) Plasma membrane-mediated effects of thyroid hormones. In Cumming, I. A., Funder, J. W., Mendelsohn, F. A. O. (eds) *Endocrinology 1980*. Canberra: Australian Academy of Sciences.

Valenta, L. J., Florsheim, W. C., Sharma, B. S. (1982) Acute effects of iodine on the stimulated rat thyroid. *Endocrinology*, **111**, 1721.

Werner, S. C., Ingbar, S. H. (eds) 1978 *The Thyroid*, 4th eds. New York: Harper & Row.

Multiple choice questions

1. TSH acts on the thyroid to increase:
  1. iodide trapping;
  2. thyroxine synthesis;
  3. thyroglobulin synthesis and dissolution (breakdown);
  4. triiodothyronine release into blood;
  5. calcitonin secretion;
  6. adenyl cyclase activity;
  7. thyroxine binding protein release into blood

2. In untreated thyrotoxicosis:
  1. serum TSH levels are usually elevated;
  2. alterations of humoral immunity (B cell function) are often demonstrable;
  3. alterations of cell-mediated immunity (T cell function) are often demonstrable;
  4. immunoglobulins capable of stimulating human thyroid tissue and not guinea-pig or mouse thyroid are sometimes found;
  5. elevated levels of thyroxine binding globulin (TBG) are diagnostic of the disease;
  6. administration of T4 or T3 fails to suppress the thyroid overactivity.

3. In the thyroid:
  1. tyrosine molecules which are free in the follicle lumen are iodinated to form MIT and/or DIT;
  2. pinocytic vacuoles of thyroglobulin coalesce with lysosomes in the cell cytoplasm and proteolysis releases T3 and T4;
  3. pseudohalides (e.g. SCN^-) competitively inhibit I^- uptake, I^- administration may overcome this inhibition;
  4. following thiocarbamide administration the T/S concentration ratio for I^- increases;
  5. LATS-protector is a thyroid auto-antibody present in thyrotoxic plasma and is a human thyroid stimulator;
  6. a hormone is produced which raises the plasma calcium level;
  7. an iodide pump operates in association with a Na^+-K^+ dependent ATP-ase system;
  8. when deiodinase is absent, abnormal quantities of MIT and DIT appear in the plasma.

4. In thyrotoxicosis:
  1. propylthiouracil lowers serum T3 levels partly by blocking peripheral conversion from T4;
  2. symptoms of sympathetic overactivity are due to raised catecholamine levels;
  3. carbimazole blocks hormone synthesis at more than one site in the biosynthetic pathway;

4. one drop of Lugol's solution daily will lower the BMR;
5. iodine is used concurrently in the final preparation of the patient treated with perchlorate;
6. elevated TSH levels can be observed in treated patients;
7. occurring in pregnancy the fetal thyroid is suppressed by maternal hormone excess.

5. Thyroxine (T4) and triiodothyronine (T3):
1. circulate in plasma predominantly bound to proteins;
2. levels increase in plasma during pregnancy and mild hyperthyroidism normally occurs;
3. secretion decreases following major surgery;
4. have no effect on erythropoiesis;
5. at physiological levels uncouple oxidative phosphorylation and hence heat production is increased and ATP synthesis is reduced;
6. increases the activity of the Na^{+}-pump ATP-ase and hence the consumption (and production) of ATP leading to increased heat production.

6. With respect to tests of thyroid function:
1. techniques which provide an estimate of the free (unbound) serum T4 level are currently the simplest and most reliable for screening of patients with suspected thyroid disease;
2. the serum T3 level is rarely elevated in thyrotoxicosis;
3. measurement of T3-resin uptake alone is a reliable test of thyroid function;
4. failure of TSH response to the intravenous administration of thyrotrophin releasing factor (TRF) provides confirmatory evidence to support a clinical diagnosis of thyrotoxicosis;
5. the basal serum TSH level is a useful test in the diagnosis of primary hypothyroidism;
6. the basal serum TSH level is a useful test in the diagnosis of thyrotoxicosis;
7. the administration of a cough mixture containing iodide does not interfere with serum T4 measurement.

7. Inorganic iodide:
1. is actively transported into cells in salivary glands and becomes incorporated into T3 and T4;
2. concentration in plasma is 0.3 to 0.7 μg/100 ml;
3. uptake by thyroid is about 70 μg/day, which is approximately the amount required for the synthesis of 100 μg/day of T4;
4. in the thyroid accounts for only about 2% of the total iodine in the gland, the rest being organically bound;
5. intake should not be less than 20 to 30 μg/day in order to maintain euthyroidism;

6. can be replaced by chloride during adaptation to iodide deficiency and euthyroidism is maintained;
7. concentration ratio between the thyroid and plasma is normally about 25 (i.e. T/P = 25);
8. uptake by thyroid cells is competitively inhibited by thiocarbamides.

8. Graves' disease:
1. diffuse thyroid hyperplasia is a cardinal feature;
2. solitary thyroid adenoma is a cardinal feature;
3. is associated with high TSH levels;
4. is associated with high T3 levels;
5. may not have elevated T4 levels;
6. is not the only form of hyperthyroidism.

9. In the thyroid:
1. following deiodination of MIT and DIT in the cell cytoplasm most of the inorganic iodide recirculates to the colloid and becomes organically bound;
2. following SCN^- administration the T/S concentration ratio for I^- remains unchanged;
3. about equal amounts of T3 and T4 are formed daily;
4. normally only very small amounts of MIT and DIT (< 1 µg/100 ml plasma) are released into the blood and these iodotyrosines are physiologically inactive;
5. the intrinsic ability to secrete small quantities of iodothyronines is retained following hypophysectomy;
6. an early response to TSH is proteolysis of thyroglobulin.

10. With respect to hypothyroidism:
1. more than 50% of patients with hypothyroidism have associated defects of parathormone secretion;
2. severe hypothyroidism can be induced by excessive administration of anti-thyroid drugs;
3. the serum cholesterol level is of value in making a differential diagnosis between primary and secondary hypothyroidism;
4. in primary hypothyroidism, the skin is usually thickened, coarse and cold;
5. the skin in secondary hypothyroidism is characteristically the same as in primary hypothyroidism;
6. TSH levels increase markedly following the intravenous administration of 200 µg of TRF in primary hypothyroidism.

11. In the thyroid:
1. iodide enters cells by passive diffusion;
2. iodide enters cells against a concentration gradient;
3. iodide enters cells against an electric gradient;

4. hormone synthesis takes place at the cell-colloid interface;
5. thiocyanate (or perchlorate) competitively inhibits iodide uptake;
6. thiocarbamides (e.g. thiouracil) inhibit oxidation of I^- to I_2 and the synthesis of thyroxine;
7. LATS (long acting thyroid stimulator) resembles TSH in its actions;
8. daily synthesis of triiodothyronine is greater than that of thyroxine;
9. thyroxine synthesis is increased on exposure of a man to 38 °C for more than 48 hours.

12. T4 and T3:
1. are conjugated with glucuronic acid in the liver;
2. changes in thyroid hormone secretion are essential in acute thermogenic response to cold in man;
3. increase the BMR of brain cells in the adult;
4. increase Na^+ and K^+ fluxes in responsive tissues;
5. have no effect in mitochondria;
6. during starvation for 48 hours increase the mobilisation of FFA from fat cells;
7. increase the peripheral resistance in the circulation in man;
8. act via adenyl cyclase;
9. increase rate of glucose absorption from small intestine by increasing Na^+-K^+ ATP-ase activity.

13. High plasma inorganic iodide in Graves' disease causes:
1. depressed radioactive iodine uptake;
2. depressed organification of iodine;
3. depressed absolute iodine uptake;
4. increased thyroidal iodine clearance;
5. increased T3 secretion;
6. decreased hormone release from the gland.

14. With respect to the action of thyroid hormones on the BMR, which is the odd one out?
1. liver;
2. kidney;
3. heart;
4. muscle (skeletal);
5. anterior pituitary;
6. vascular smooth muscle.

15. Which of the following are true statements?
1. in assessing thyroid function biochemically, it is sufficient to measure total T4 level;
2. when α-2 globulin levels are raised, the T3-resin uptake is reduced;

3. in women taking oral contraceptives T3-resin uptake is elevated;
4. due to increase in dietary iodine, the normal range for ^{131}I uptake is falling;
5. in the T3-suppression test for thyrotoxicosis, ^{131}I uptake is measured before and after five days of T3 20 to 50 μg 8 hourly;
6. if thyroid function is normal, the T3 suppression test results in a 5 to 10% reduction in ^{131}I uptake;
7. in hyperthyroidism associated with exophthalmos, a fall in thyroid hormone level is followed by improvement in the eye disease;
8. as thyroid function falls following ^{131}I treatment of thyrotoxicosis, clinical features precede changes in thyroid hormone level by 1 to 2 weeks.

16. Endocrine abnormalities associated with cirrhosis of the liver include:
1. a high incidence of diabetes mellitus;
2. abnormal thyroxine binding;
3. raised aldosterone levels due to impaired aldosterone clearance;
4. raised prolactin levels;
5. raised thyroxine levels due to diminished hormone conjugation;
6. gonadal atrophy due to impaired gonadotrophin secretion.

Answers

1. 1, 2, 3, 4, 6
2. 2, 3, 4, 6
3. 2, 3, 4, 5, 7, 8
4. 1, 3, 4, 6
5. 1, 3, 6
6. 1, 4, 5, 7
7. 2, 3, 4, 5, 7
8. 1, 4, 5, 6
9. 1, 4, 5, 6
10. 2, 3, 4, 6
11. 2, 3, 4, 5, 6, 7
12. 1, 4, 6, 9
13. 1, 2, 3, 6
14. 5
15. 2, 4, 5
16. 1, 2

---

4. Adrenal cortex

STRUCTURE ACTIVITY RELATIONSHIP OF STEROID HORMONES

The basic structure for all steroid hormones is the cyclopentanoperhydrophenanthrene nucleus.

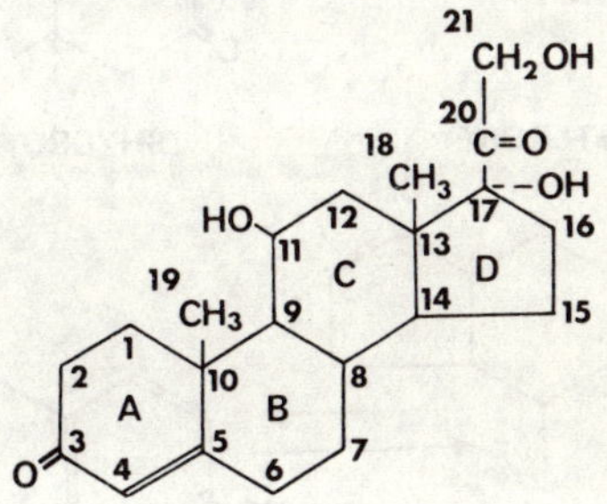

Fig. 4.1 Cortisol

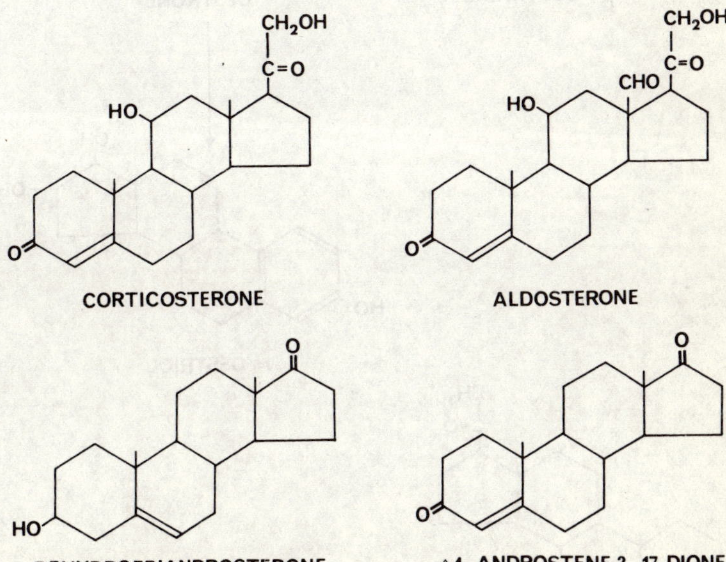

Fig. 4.2 Structural formulae of some adrenal steroids

Steroids with 18 carbon atoms (C18 steroids) are oestrogens, those with 19 are androgens (C19 steroids).

The 21 carbon structure with a ketone at position 20 is necessary for glucocorticoid, mineralocorticoid and progestational activities. An α OH at position 17 imparts maximal glucocorticoid activity. The combination of an aldehyde at position 18 and an OH at position 11

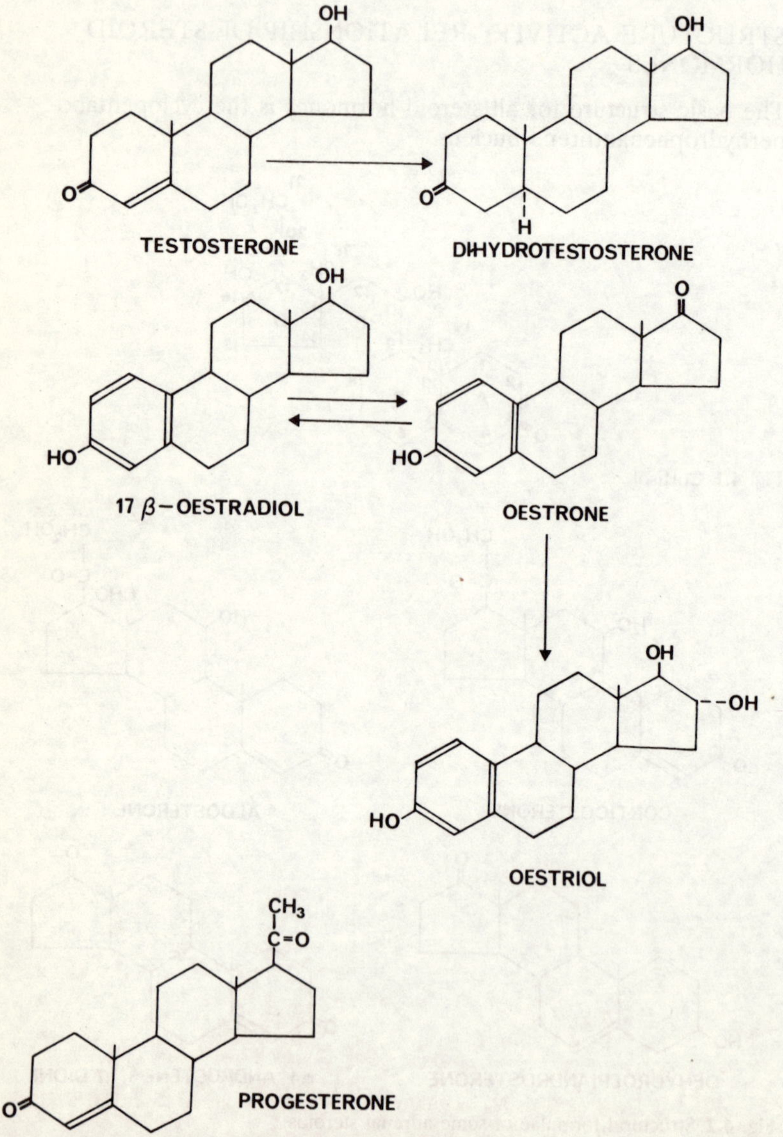

TESTOSTERONE → DIHYDROTESTOSTERONE

17 β−OESTRADIOL ⇄ OESTRONE → OESTRIOL

PROGESTERONE

Fig. 4.3 Structural formulae of some testis and ovary steroids

is responsible for the marked mineralocorticoid activity of aldosterone.

In synthetic steroids 9α-fluorine substitution intensifies both the glucocorticoid and mineralocorticoid activities.

— β OH = OH lies above plane of ring
- - - α OH = OH lies below plane of ring

STEROID SYNTHESIS IN ADRENAL CORTEX, TESTIS AND OVARY

This is shown in Figure 4.4.

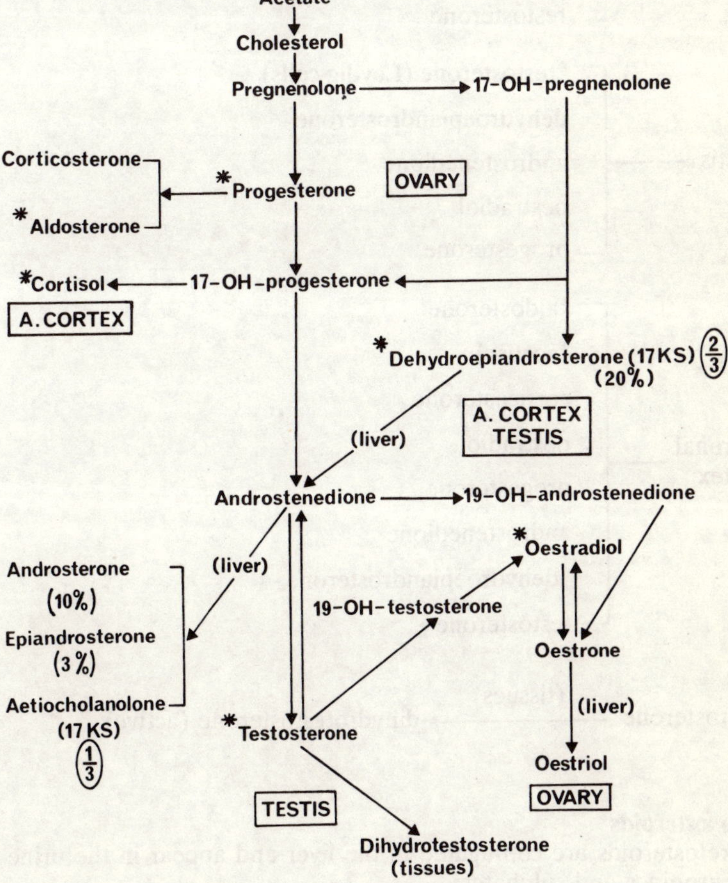

Fig. 4.4 Steroid synthesis in adrenal cortex, testis and ovary

Key: 17KS = 17-oxosteroids (17-ketosteroids) (⅔ derived from adrenal cortex; ⅓ from testis)
% = androgenic activity with respect to testosterone
* = main hormones secreted

Steroidogenic pathway determined by enzyme profile in gland cells. In testis and ovary 21 and 11 hydroxylases are absent, hence no synthesis of gluco- or mineralo- corticoids. All steroidogenic tissues synthesise androgens, oestrogens and progestogens, and develop from the same embryonic tissue (mesoderm of urogenital ridge).

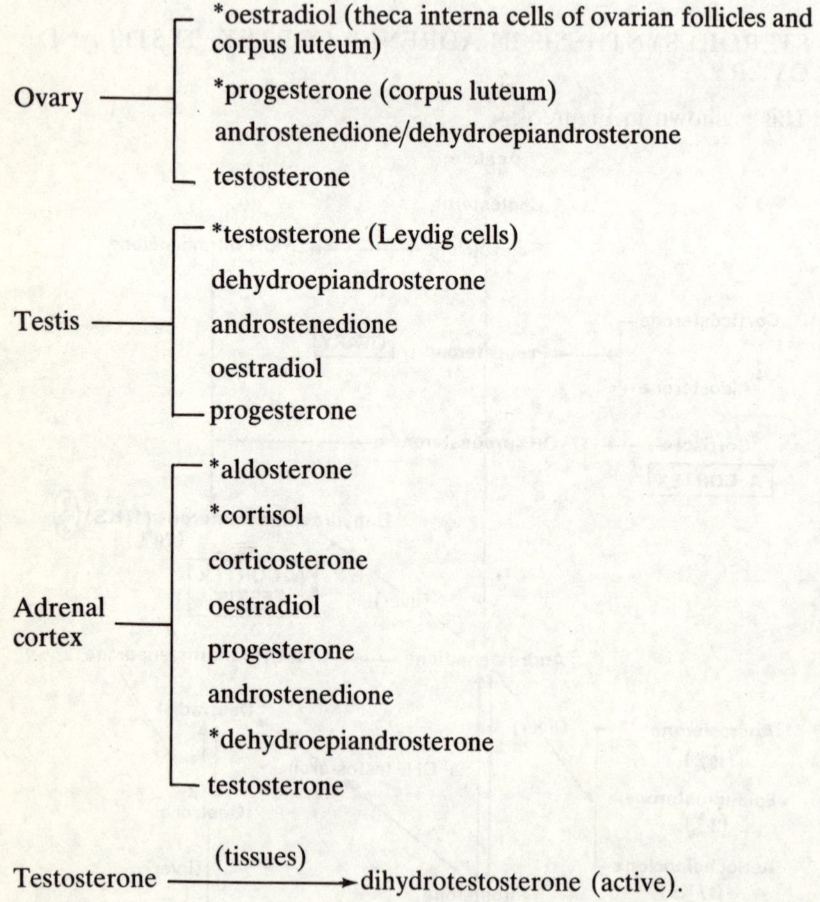

Testosterone —(tissues)→ dihydrotestosterone (active).

17-oxosteroids
17-ketosteroids are conjugated in the liver and appear in the urine as glucuronides and sulphates.

* Major hormones secreted

Hepatic metabolism of androgens to 17-oxosteroids

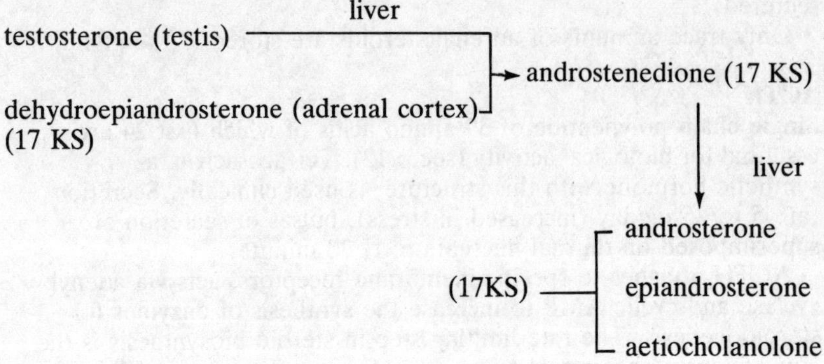

testosterone is not a 17-oxosteroid

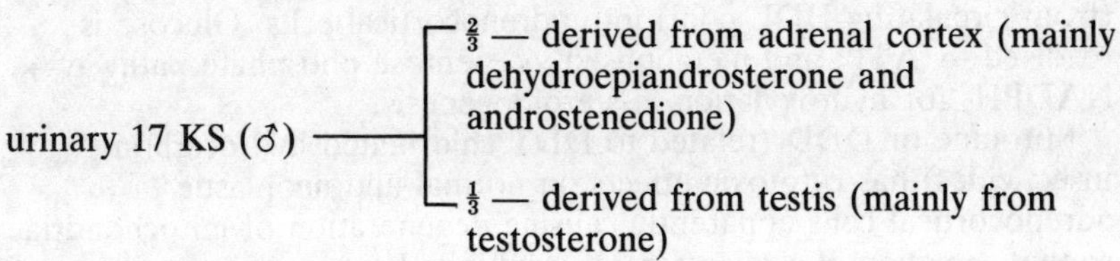

Thus 17 KS are not a good index of biological androgenic activity (testicular testosterone main contributor).

FUNCTIONAL ANATOMY OF ADRENAL CORTEX

See Figure 4.5.

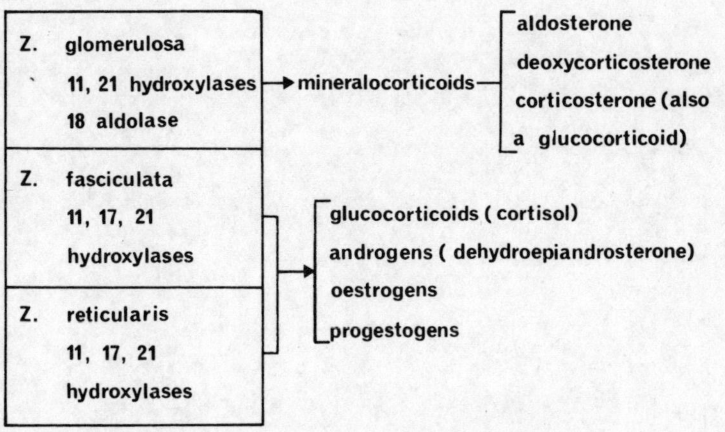

Fig. 4.5 Functional anatomy of adrenal cortex

The zona fasciculata synthesises (mainly) glucocorticoids, and zona reticularis (mainly) androgens, oestrogens and progestogens (not secreted).

Only trace amounts of adrenal steroids are stored in the gland.

ACTH

Single chain polypeptide of 39 amino acids of which first 24 are essential for biological activity (see p.12). Tetracosactrin, a synthetic hormone with this structure, is used clinically. Secretion rate 5 to 25 μg/day (increased in stress); pulses of secretion are superimposed on diurnal fluctuation; $t\frac{1}{2}$ 20 minutes.

ACTH attaches to specific membrane receptors, acts via adenyl cyclase and cyclic AMP to increase the synthesis of enzymes for steroidogenesis. The rate limiting step in steroid biosynthesis is the conversion of cholesterol to pregnenolone, and the rate of this conversion is greatly increased by ACTH (see Fig. 4.6). ACTH (in presence of Ca^{++}) stimulates the entry of glucose and cholesterol (from circulating HDL pool) into adrenocortical cells. Glucose is oxidised → ATP; and metabolised via pentose phosphate pathway→ $NADPH_2$ for hydroxylation in steroidogenesis.

Mitotane or DDD (related to DDT chlorinated hydrocarbon insecticides) has cytotoxic effects on normal and neoplastic adrenocortical cells apparently causing degeneration of mitochondria so that secretion of corticosteroids is inhibited.

SYNTHESIS OF ADRENAL STEROIDS

This is shown in Figure 4.6.

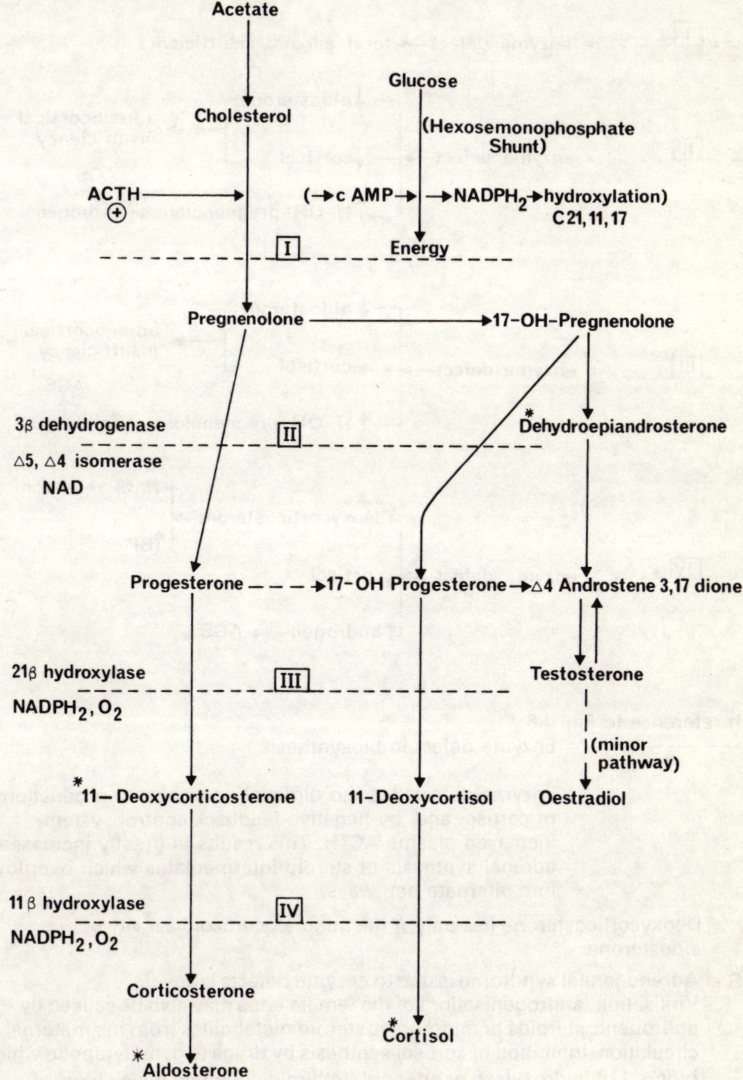

Fig. 4.6 Synthesis of adrenal steroids

Key: * Major hormones secreted — degraded to inactive tetrahydro forms in liver, conjugated with glucuronic acid → urine

⊕ Aldosterone is produced in zona glomerulosa only, and angiotensin II also activates the pathway from cholesterol

ENZYME DEFECTS IN BIOSYNTHESIS

See Figure 4.6.

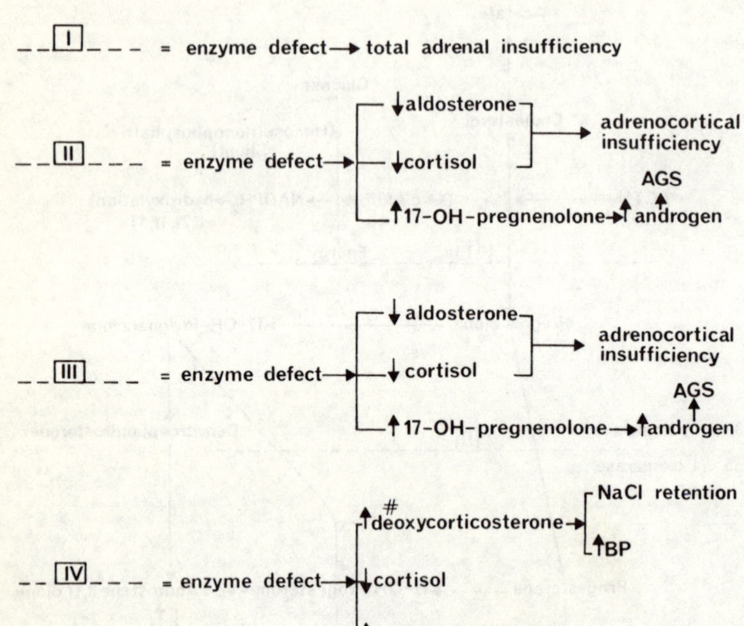

With reference to Fig. 4.6:

— — — — — — — — = Enzyme defect in biosynthesis

Enzyme defects lead to diminished or absent production of cortisol and, by negative feedback control system, increased plasma ACTH. This results in greatly increased adrenal synthesis of steroid intermediates which overflow into alternate pathways.

Key: # Deoxycorticosterone has only $\frac{1}{32}$ the mineralocorticoid activity of aldosterone.

AGS = Adrenogenital syndrome is due to enzyme defects (above). Virilisation (androgenisation) of the female fetus may also be caused by androgenic steroids or androgenic steroid metabolites from the maternal circulation. Inhibition of cortisol synthesis by drugs (e.g. metyrapone which blocks 11β hydroxylase or aminoglutethimide (Elipten) which blocks conversion of cholesterol to Δ^5 pregnenolone) also leads to enhanced ACTH secretion

CORTICOSTEROID BLOOD LEVELS, SECRETION RATES AND URINARY EXCRETION VALUES

Blood level
- cortisol 5–25 μg/100 ml (140–700 nmol/l)
- corticosterone 1–1.5 μg/100 ml (35–50 nmol/l)
- aldosterone 0.004–0.008 μg/100 ml (100 – 200 pmol/l)

Secretion rate/day
- cortisol 15–25 mg
- aldosterone 0.05–0.2 mg
- deoxycorticosterone 2–3 mg
- dehydroepiandrosterone 15–25 mg

Excretion/day
- 17-oxosteroids
  - 6–20 mg (20–70 μmol) (men)
  - 6–15 mg (20–50 μmol) (women)
- 17-oxogenic steroids (17-OH-corticosteroids) 4–16 mg (15– 55 μmol)

Cortisol $t_{\frac{1}{2}}$ ~60 minutes; aldosterone $t_{\frac{1}{2}}$ ~20 minutes. Aldosterone blood level increases in upright position and when ambulant. Cortisol and corticosterone bind to corticosteroid binding globulin (CBG) or transcortin; ~ 10 per cent is free in plasma. Aldosterone binds to plasma proteins; ~ 40% is free in plasma. 17-oxogenic steroids are the degradation products of adrenocortical steroids with an OH at C17.

Diurnal rhythm in ACTH secretion — highest about 8 a.m. (10 to 80 pg/ml or 2 to 18 pmol/l), lowest around midnight (<10 pg/ml or <2 pmol/l), with parallel changes in cortisol secretion. Circadian rhythm is entrained to the dark/light cycle.

This rhythm is absent in all patients with Cushing's syndrome due to an adrenal tumour. However, the rhythm is present in most cases of Cushing's syndrome due to excessive production of ACTH, which is the commonest variety in adults.

Diurnal rhythm is also absent in patients with specific lesions in limbic-hypothalamus system, during stress, in blind people and is abolished by some drugs.

RELATIVE ACTIVITIES OF NATURAL AND SYNTHETIC CORTICOSTEROIDS

See Table 4.1.

Table 4.1 Relative activities of natural and synthetic corticosteroids (V. Cortisol)

Steroid	Glucorticoid activity	Mineralocorticoid activity
Cortisol (hydrocortisone)	1	1
Prednisolone (△′ cortisol)	5	1
Triamcinolone (9 α-fluoro- 16∝ OH-prednisolone)	5	0
Dexamethasone (9 α-fluoro- 16∝ CH_3-prednisolone)	25	0
Betamethasone (9 α-fluoro-16β CH_3-prednisolone)	25	0
Aldosterone	0	300
Fludrocortisone (9 α-fluorocortisol)	10	125
Deoxycorticosterone	0	10

MECHANISM OF ACTION OF ALDOSTERONE

See Figure 4.7.

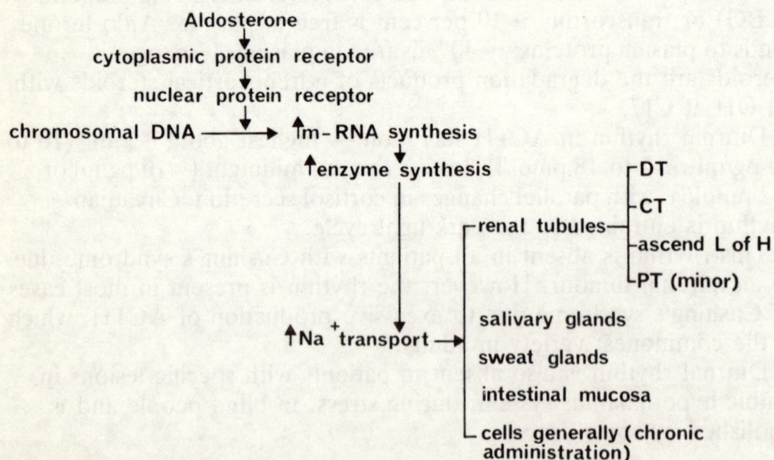

Fig. 4.7 Mechanism of action of aldosterone
Key: DT = distal tubule; CT = collecting tubule; L of H = loop of Henle; PT = proximal tubule

ACTIONS OF ALDOSTERONE

These are outlined in Figure 4.8.

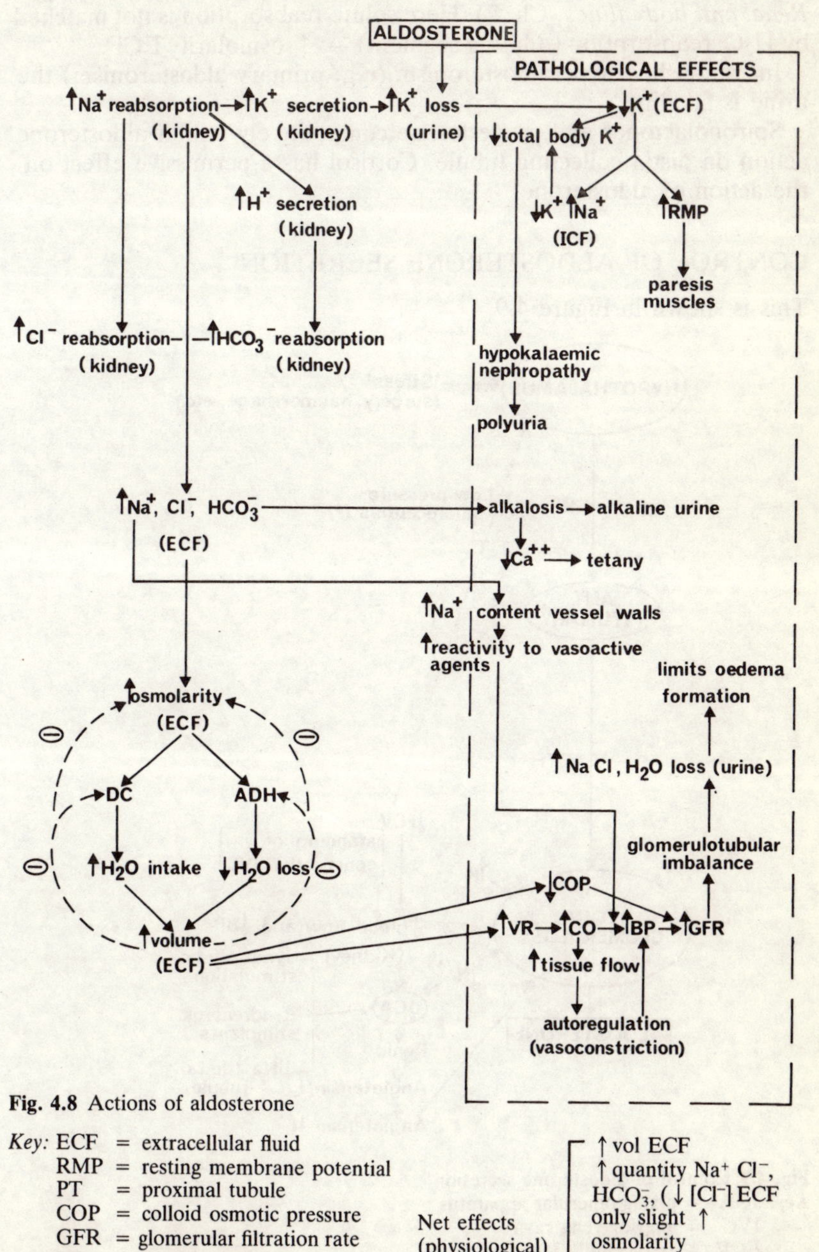

Fig. 4.8 Actions of aldosterone

Key: ECF = extracellular fluid
RMP = resting membrane potential
PT = proximal tubule
COP = colloid osmotic pressure
GFR = glomerular filtration rate
VR = venous return
CO = cardiac output
DC = drinking centre

Net effects (physiological):
- ↑ vol ECF
- ↑ quantity Na^+ Cl^-, HCO_3^-, (↓ [Cl^-] ECF
- only slight ↑ osmolarity ECF (~5 mOsm/l) and [Na^+]
- ↓ quantity and [K] ECF

Aldosterone acting on distal tubules and collecting ducts (i.e. cation exchange region) increases exchange of K^+ and H^+ (excreted) for Na^+ (reabsorbed) (see Hawker, *Notebook of Medical Physiology: Renal and body fluids*, Ch. 2). Here solute reabsorption is not matched by H_2O reabsorption (diluting segment) → ↑ osmolarity ECF.

In established hyperaldosteronism (e.g. primary aldosteronism) the urine is alkaline.

Spironolactones and progesterone competitively inhibit aldosterone action on distal collecting tubule. Cortisol has a permissive effect on the action of aldosterone.

CONTROL OF ALDOSTERONE SECRETION

This is shown in Figure 4.9.

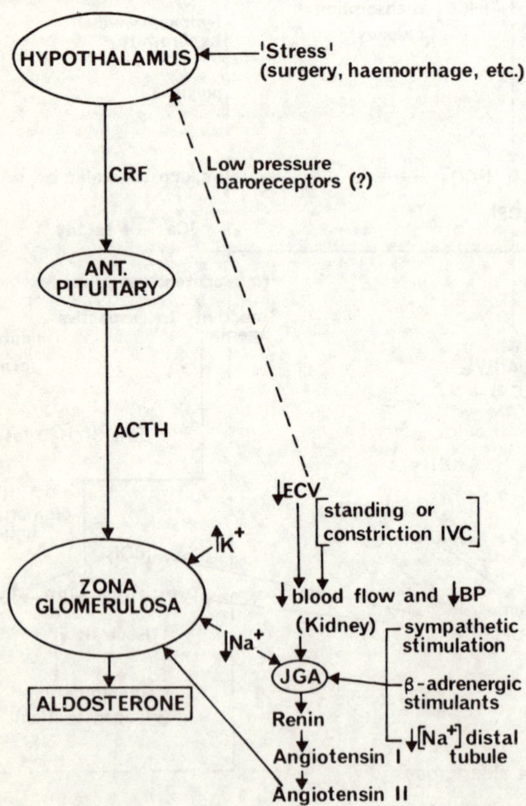

Fig. 4.9 Control of aldosterone secretion
Key: JGA = juxtaglomerular apparatus
IVC = inferior vena cava
ECV = extracellular volume

Angiotensin II → angiotensin III. Angiotensin II stimulates production of aldosterone, and is rapidly inactivated. Angiotensin III is slowly inactivated, and has some residual effect on aldosterone secretion.

At physiological levels ACTH stimulates aldosterone secretion, and ACTH administration increases aldosterone production (and enhances angiotensin response).

Basal secretion of aldosterone occurs in absence of anterior pituitary.

Renin-angiotensin (cAMP mediated) mechanism is major regulating system.

In 'stress' situations, e.g. surgery, burns etc., ACTH-induced secretion of aldosterone is increased.

Aldosterone causes expansion of extracellular volume (and increases sodium) which shuts off renin-angiotensin control system (see Hawker, *Notebook of Medical Physiology: Renal and body fluids*, Ch. 7).

ACTIONS OF CORTISOL

1. Metabolic (see Fig. 4.10 and p. 85)
2. Mild aldosterone-like action
3. Permissive effect on H_2O diuresis (in adrenal insufficiency ability to excrete a H_2O load is impaired due to ↓ GFR; cortisol ↑ GFR in this condition)
4. Permissive effect on vascular reactivity to catecholamines and stimulatory effect on myocardial contractility

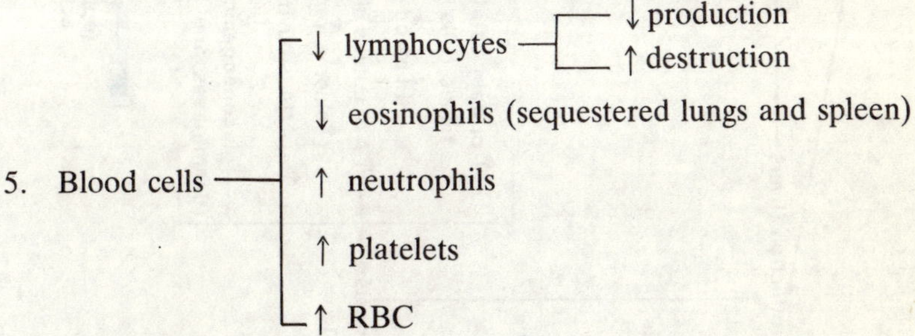

6. ↓ Inflammatory and allergic reactions (↓ histamine release from ↓ numbers of mast cells; stabilises lysosome membranes; ↓ digestive/phagocytic activity of monocytes; ↓ multiplication of plasma cells; ↓ bradykinin synthesis); ↓ immune system
7. ↑ Involution thymus and lymph nodes (eventually ↓ production antibodies)
8. ↑ Resistance to 'stress'

9*. Retarded growth and muscle wasting

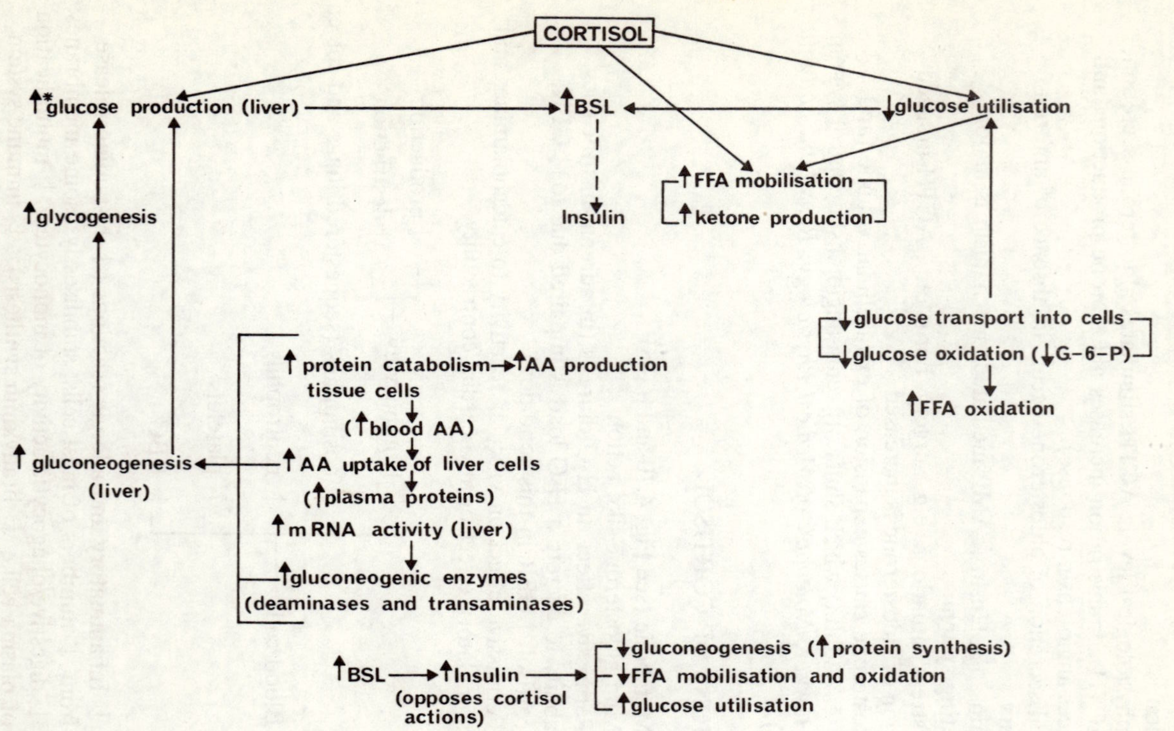

Fig. 4.10 Metabolic actions of cortisol

10*. ↓ Protein matrix in bone → osteoporosis; ↑ urinary Ca

11*. ↑ Secretion HC1 and pepsin (→ulcers)
12*. ↓ Synthesis of ground substance, and ↓ growth/development of fibroblasts → ↓ growth of connective tissue (↓ fibrosis)
13. Role in parturition (see p. 226).

Glucocorticoids exert effects on brain neurones and maintain an internal environment for normal cerebral activity (sensory and higher cortical function). These functions become disturbed when the circadian rhythm of cortisol secretion is upset e.g. during jet-lag.

CONTROL OF CORTISOL SECRETION

This is shown in Figure 4.11.

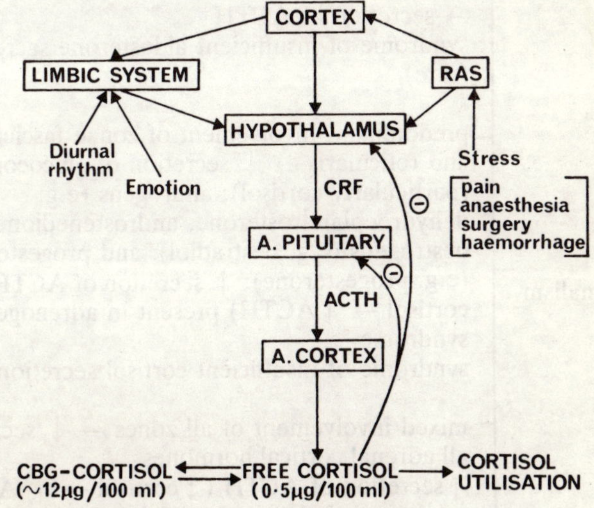

Fig. 4.11 Control of cortisol secretion

Key: CBG = corticosteroid binding globulin (transcortin)
RAS = reticular activating system

* Not at physiological levels.

Ach (5HT, ADH) → ↑ CRF; NA and GABA → ↓ CRF.

The plasma concentration of free cortisol (which is in equilibrium with cortisol bound to CBG) is regulated by the negative feedback servo-system.

In conditions where the concentration of CBG is increased (e.g. pregnancy, oestrogen administration) the total plasma cortisol will be increased but the free cortisol (the biologically active form, which is the regulator of the servo-system) is held constant. Hence pregnancy is a eucorticoid state despite elevated total plasma cortisol.

DISORDERS OF ADRENOCORTICAL FUNCTION

Hypoadrenalism

1. Primary hypoadrenalism

Primary hypoadrenalism

- initial abnormality is in adrenal cortex
- predominant involvement of zona glomerulosa → ↓ secretion of mineralocorticoids (particularly aldosterone)
  ↔ secretion of ACTH
  syndrome of insufficient aldosterone secretion (rare)
- predominant involvement of zonae fasciculata and reticularis → ↓ secretion of glucocorticoids (particularly cortisol); androgens (e.g. dehydroepiandrosterone, androstenedione); oestrogens (e.g. oestradiol); and progestogens (e.g. progesterone); ↑ secretion of ACTH (↓ cortisol → ↑ ACTH) present in adrenogenital syndrome
  syndrome of insufficient cortisol secretion
- mixed involvement of all zones → ↓ secretion of all adrenal cortical hormones
  ↑ secretion of ACTH (↓ cortisol → ↑ ACTH)
  syndrome of chronic insufficiency of both aldosterone and cortisol secretion (Addison's disease)
- caused by autoimmune adrenalitis (idiopathic), tuberculosis, bilateral adrenalectomy, carcinomatosis, haemochromatosis

Note: increased secretion of ACTH causes pigmentation of skin.

2. Secondary hypoadrenalism

a. pituitary hypoadrenalism

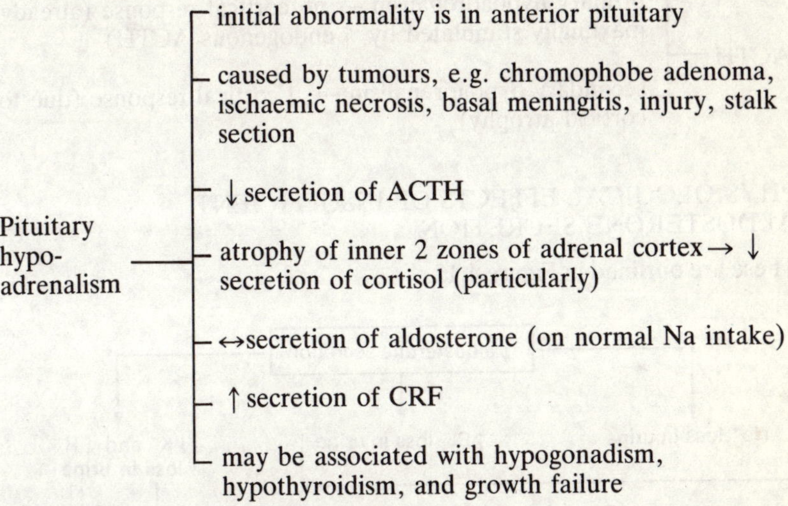

b. hypothalamic hypoadrenalism

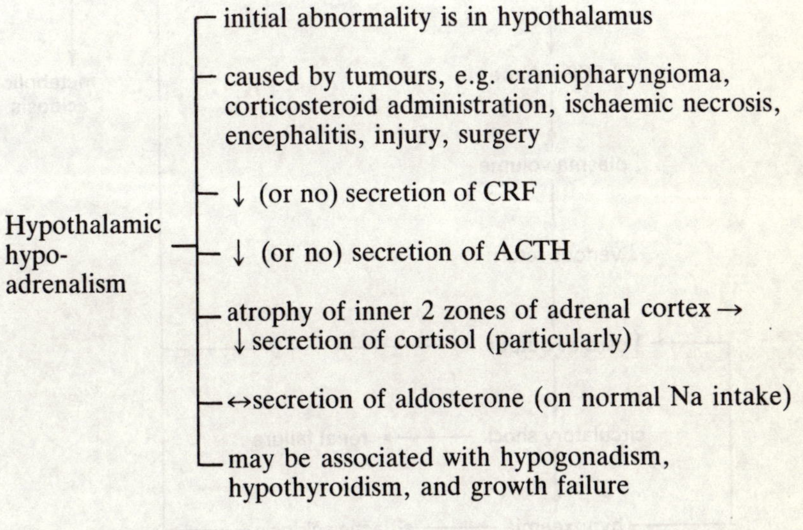

Note
- in (a) and (b) since aldosterone secretion is normal there is little (or no) disturbance of H_2O/electrolyte metabolism
- in (a) and (b) since ACTH secretion is reduced there is no pigmentation of skin

A test dose of ACTH distinguishes between primary and secondary hypoadrenalism:

ACTH —
- primary hypoadrenalism → no cortical response (already maximally stimulated by ↑ endogenous ACTH)
- secondary hypoadrenalism → ↓ cortical response (due to cortical atrophy)

PHYSIOLOGICAL EFFECTS OF INSUFFICIENT ALDOSTERONE SECRETION

These are outlined in Figure 4.12.

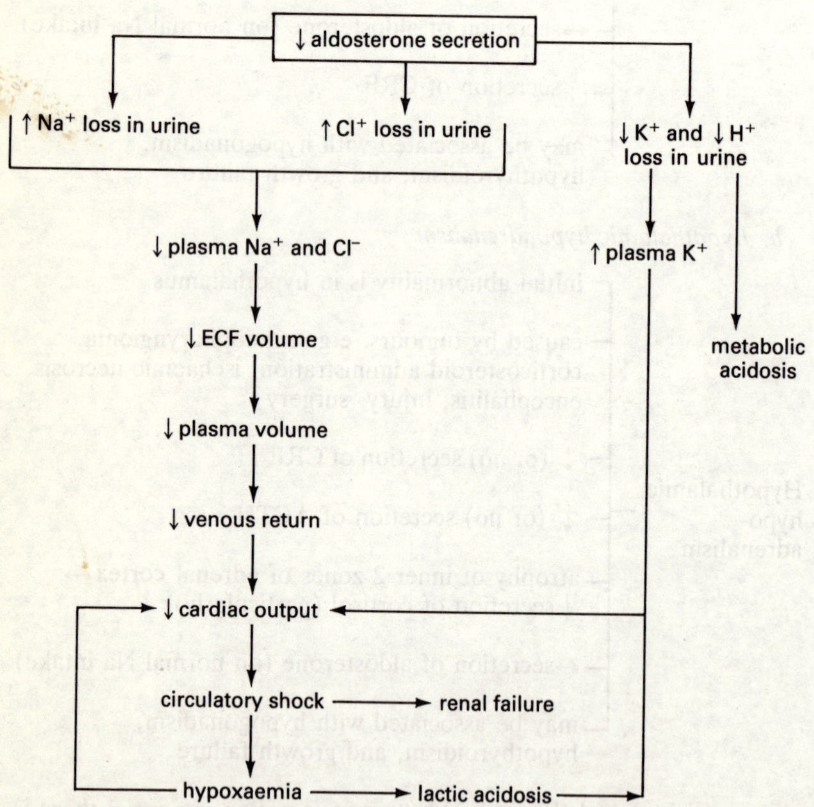

Fig. 4.12 Physiological effects of insufficient aldosterone secretion

↑ NaC1 loss in urine → ↑ water loss in urine (↑ urine flow)

PHYSIOLOGICAL EFFECTS OF INSUFFICIENT CORTISOL SECRETION

a. Metabolic effects

The metabolic effects are outlined in Figure 4.13.

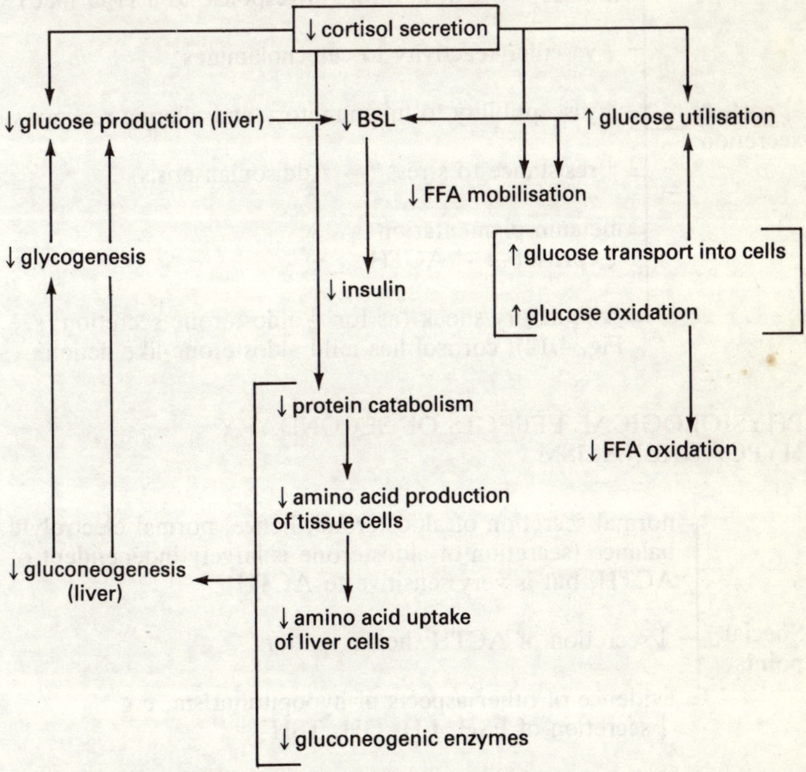

↓ cortisol secretion

↓ glucose production (liver) → ↓ BSL ← ↑ glucose utilisation

↓ FFA mobilisation

↓ glycogenesis

↑ glucose transport into cells

↓ insulin

↑ glucose oxidation

↓ protein catabolism

↓ FFA oxidation

↓ amino acid production of tissue cells

↓ gluconeogenesis (liver)

↓ amino acid uptake of liver cells

↓ gluconeogenic enzymes

Fig. 4.13 Metabolic effects of insufficient secretion of cortisol

After a carbohydrate meal a reactive hypoglycaemia may occur:
↑ BSL → ↑ insulin → reactive hypoglycaemia

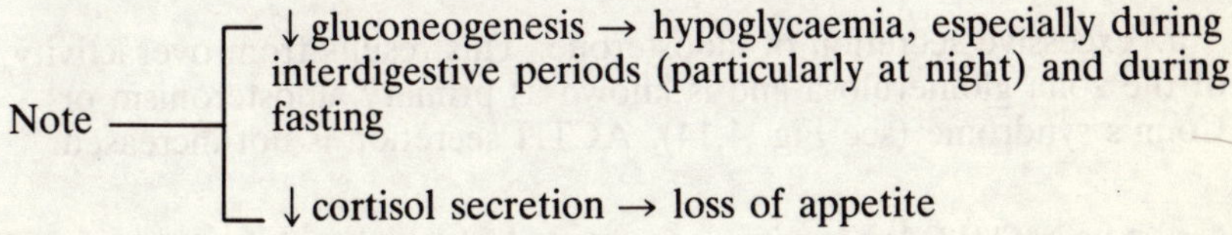

Note — ↓ gluconeogenesis → hypoglycaemia, especially during interdigestive periods (particularly at night) and during fasting

— ↓ cortisol secretion → loss of appetite

b. Other effects

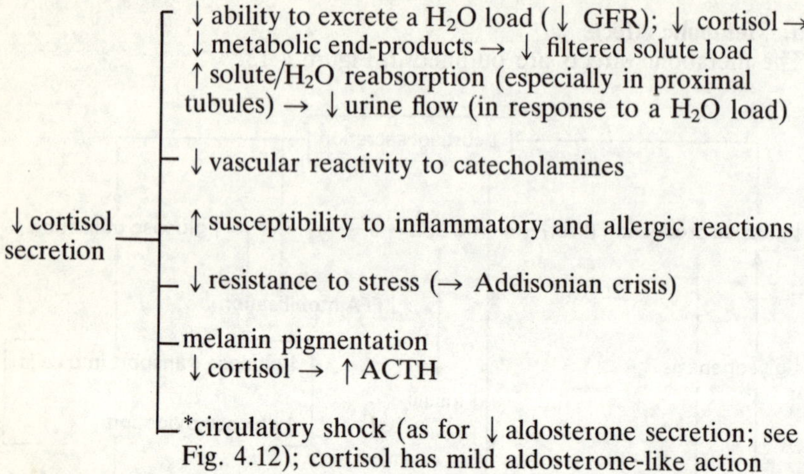

PHYSIOLOGICAL EFFECTS OF SECONDARY HYPOADRENALISM

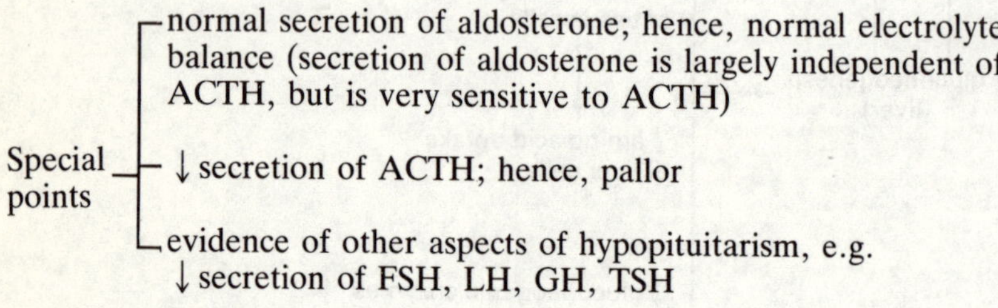

Adrenogenital syndrome

See page 74.

Hyperadrenalism

1. Primary hyperadrenalism

a. excessive secretion of aldosterone. This results from overactivity of the zona glomerulosa and is known as primary aldosteronism or Conn's syndrome (see Fig. 4.14). ACTH secretion is not increased.

* = severe NaCl/H_2O loss in urine → circulatory shock (and hyperkalaemia); treatment with saline solutions and/or aldosterone is not enough: glucocorticoid therapy is essential

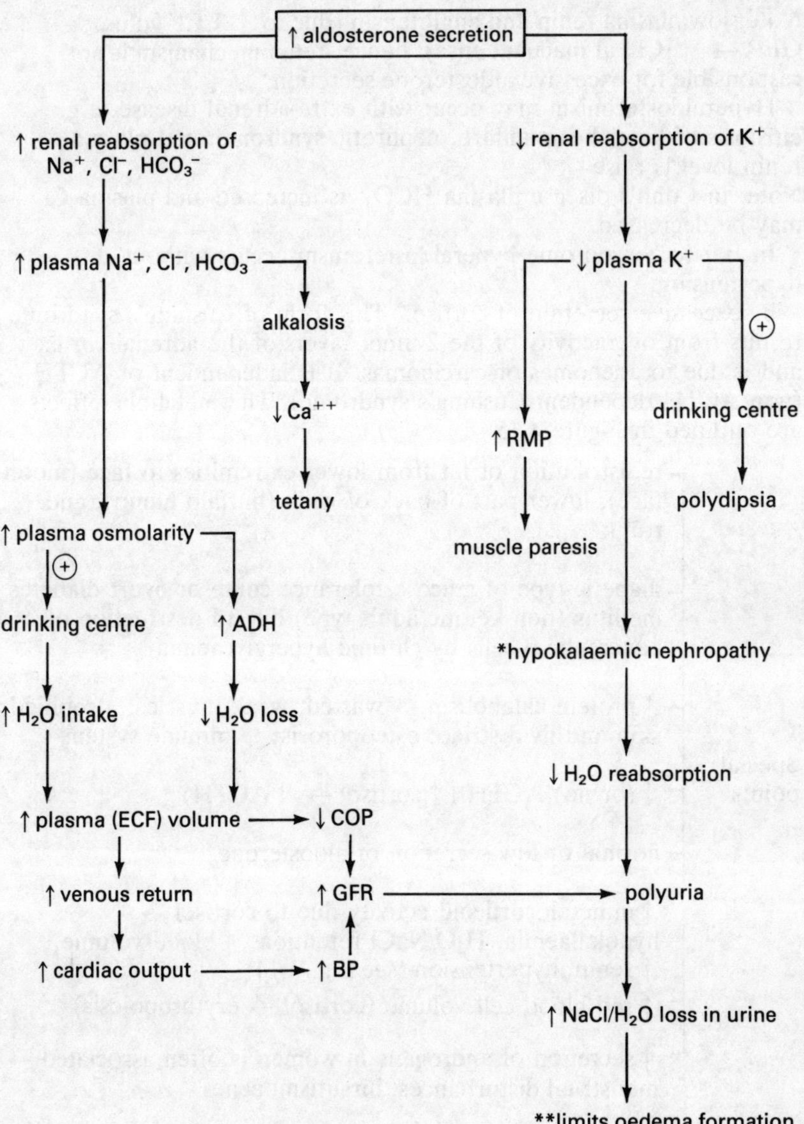

Fig. 4.14 Effects of excessive secretion of aldosterone on water and electrolyte metabolism
RMP = resting membrane potential
* = hypokalaemic nephropathy reduces ADH effect on distal/collecting tubules
** = oedema is most unusual (in absence of heart failure) and, when present, precedes hypokalaemic nephropathy

Note: low plasma renin and angiotensin (due to ↑ ECF volume → ↑ GFR → ↑ [Cl^-] at macula densa); hence, renin mechanism is not responsible for excessive aldosterone secretion.

Hyperaldosteronism may occur with extra-adrenal disease, e.g. cirrhosis of liver, heart failure, nephrotic syndrome; and plasma renin level is raised.

Note: in Conn's disease plasma HCO_3^- is increased and plasma Cl^- may be decreased.

In Bartter's syndrome hyperaldosteronism occurs without hypertension.

b. excessive secretion of cortisol. This form of Cushing's syndrome results from overactivity of the 2 inner layers of the adrenal cortex and is due to adenomas or carcinomas. It is independent of ACTH (non-ACTH dependent Cushing's syndrome). The metabolic effects are outlined in Figure 4.15.

Special points:

- redistribution of fat from lower extremities to face (moon face), lower part of back of neck (buffalo hump), and trunk
- diabetic type of glucose tolerance curve or overt diabetes mellitus (non-ketotic adult type) due to destruction of pancreatic β-cells by chronic hyperglycaemia
- ↑ protein catabolism → wasted, weak muscles; atrophied skin and livid striae; osteoporosis; ↓ immune system
- ↓ (or no) ACTH (↑ cortisol → ↓ ACTH)
- normal or low secretion of aldosterone
- ↑ mineralocorticoid activity due to cortisol → hypokalaemia, H_2O/NaCl retention, ↑ blood volume, ↓ renin, hypertension (see Fig. 4.14)
- ↑ red blood cell volume (cortisol $\xrightarrow{\oplus}$ erythropoiesis)
- ↑ secretion of androgens in women is often associated → menstrual disturbances, hirsutism, acne

Note:

- excessive secretion of cortisol by a tumour in one adrenal causes atrophy of the other, because of cortisol suppression of ACTH secretion
- adrenocortical carcinoma is commonest cause of Cushing's syndrome in children

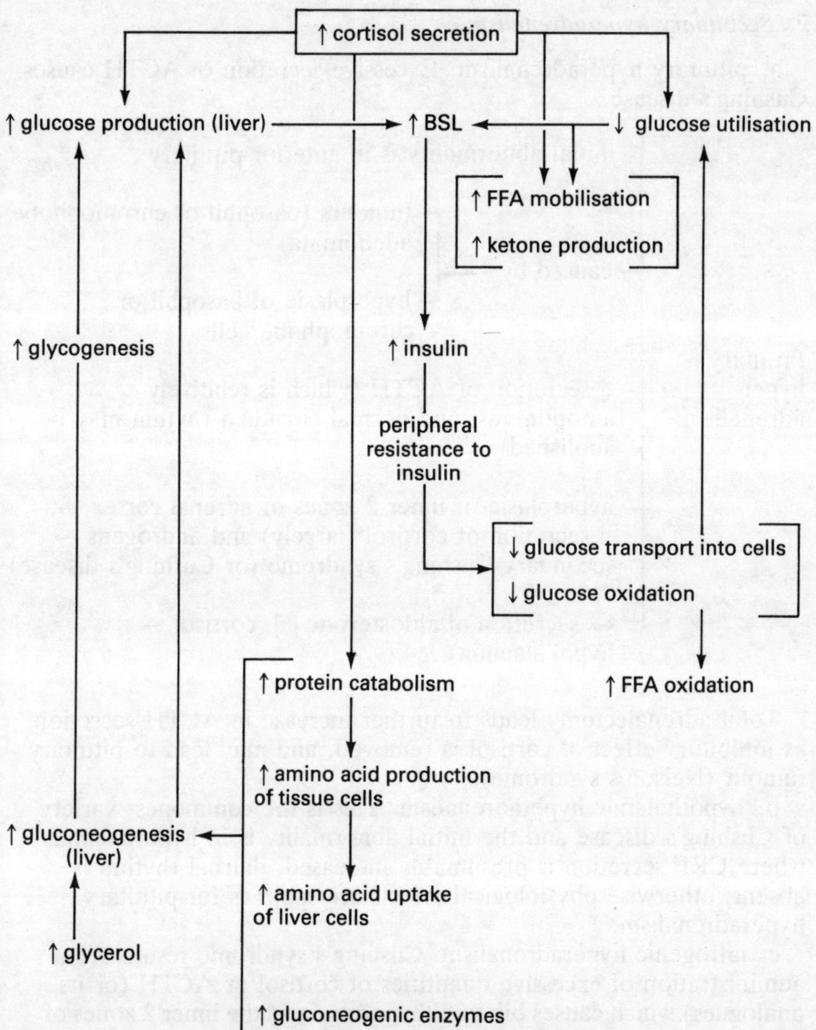

Fig. 4.15 Metabolic effects of excessive secretion of cortisol

c. excessive secretion of androgens. The effects depend on the time of excess secretion. Masculinisation occurs in the female fetus (adrenogenital syndrome, female pseudohermaphroditism); precocious pseudopuberty in males; virilisation in women (hirsutism, acne, amenorrhoea and other menstrual disorders, temporal recession of scalp hair, deepening of voice, clitoral enlargement, increased libido).

Excessive secretion of oestrogens may induce feminisation in males.

2. Secondary hyperadrenalism

a. pituitary hyperadrenalism. Excessive secretion of ACTH causes Cushing's disease.

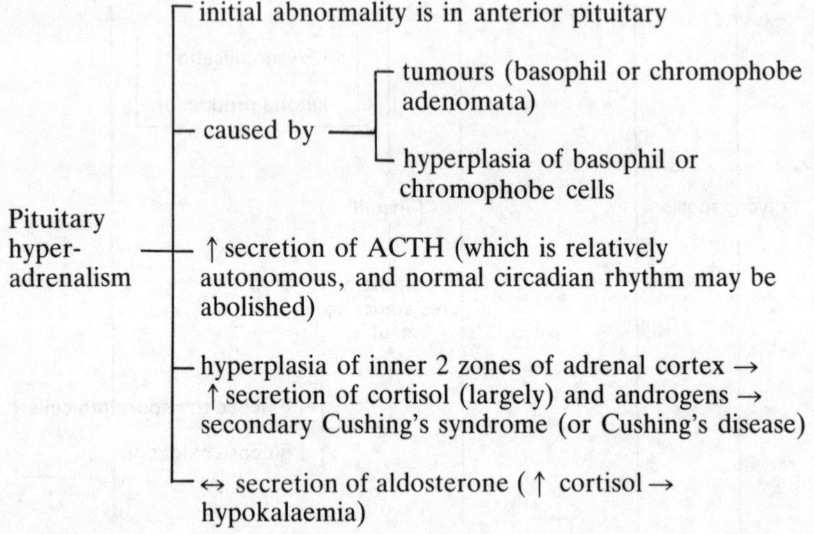

Total adrenalectomy leads to further increase in ACTH secretion as inhibitory effect of cortisol is removed, and may lead to pituitary tumour (Nelson's syndrome).

b. hypothalamic hyperadrenalism. This is the commonest variety of Cushing's disease and the initial abnormality is in hypothalamus where CRF secretion is presumably increased; diurnal rhythm is absent; otherwise physiological effects are same as for pituitary hyperadrenalism.

c. iatrogenic hyperadrenalism. Cushing's syndrome results from administration of excessive quantities of cortisol or ACTH (or its analogues) which causes bilateral hyperplasia of the inner 2 zones of the adrenal cortex.

d. ectopic ACTH hyperadrenalism. Cushing's syndrome may result from ectopic secretion by non-endocrine tumours (e.g. bronchial carcinoma) of excessive amounts of ACTH. The MSH-like activity of abnormally high amounts of ACTH leads to a brown pigmentation of the skin. Hypokalaemic alkalosis (due to very high levels of cortisol), and severe diabetes mellitus may result. Death usually occurs within weeks, and central obesity usually does not have time to develop.

Note: in secondary hyperadrenalism, increased secretion of cortisol co-exists with increased secretion of ACTH: in a and b from anterior pituitary, in c from administration, in d from non-endocrine tumour.

ADRENOCORTICAL FUNCTION TESTS (PRINCIPLES ONLY)

Dexamethasone suppression test

In normal subjects dexamethasone suppresses ACTH and cortisol secretion so that by the second day of administration levels of plasma cortisol and urinary metabolites of cortisol (17-hydroxycorticosteroids and 17-oxogenic steroids) become undetectable. In patients with Cushing's syndrome due to an adrenal tumour secreting cortisol, or due to ectopic production of ACTH, there is no response to dexamethasone. Patients with Cushing's syndrome due to pituitary over-production of ACTH usually show a partial response to the standard dosage of dexamethasone (0.5 mg, 6 hourly), which may become more pronounced at higher doses (2 mg, 6 hourly). Owing to the high potency of the synthetic steroid, effective doses do not contribute significantly to urinary steroid metabolites.

ACTH stimulation test

In normal subjects administration of ACTH or the synthetic analogue tetracosactrin (Synacthen, Cortrosyn) results in a rise in plasma cortisol and urinary metabolite excretion. In Addison's disease (primary adrenocortical insufficiency) this response is absent. In adrenocortical insufficiency due to hypopituitarism a subnormal response may be seen but in patients with adrenocortical atrophy the response may only be observed after repeated administration. In Cushing's syndrome this test may be helpful in combination with the dexamethasone suppression test in the differential diagnosis of adrenocortical adenoma (response variable), adrenocortical carcinoma (usually autonomous and unresponsive), and over-production of pituitary ACTH (normal or hyperresponsive).

Metyrapone stimulation test

A test of pituitary ACTH reserve used to diagnose hypopituitarism. Metyrapone (metopirone) inhibits 11β hydroxylase and hence decreases cortisol secretion, in normal subjects hypersecretion of ACTH results, with increased production of precursor steroids. Urinary 17-hydroxycorticosteroids, 17-oxogenic steroids and 17-oxosteroids increase significantly on the day of, and the day after, metyrapone. In hypopituitarism the pre-test level of these steroids is normal or low and, following metyrapone, either remains the same or increases slightly.

In pituitary-dependent Cushing's syndrome the urinary excretion of 17-oxogenic steroids increases markedly, and in adrenocortical tumour the values are unaltered or reduced. In adrenal Cushing's syndrome, the levels may fall.

Insulin tolerance test

Insulin-induced hypoglycaemia stimulates ACTH secretion and thus adrenocortical secretion; there is failure of ACTH response in hypopituitarism. In Addison's disease the rise in plasma ACTH does not stimulate cortisol or 17-oxosteroid secretion. The test is not done if Addison's disease is suspected since it is too dangerous.

Radioimmunoassay of plasma ACTH

Normal values are 10 to 80 pg/ml (2 to 18 pmol/l); elevated in Cushing's syndrome of pituitary origin (40 to 160 pg/ml) and in Addison's disease (>200 pg/ml); depressed (<10 pg/ml) in patients with functioning autonomous adrenal tumours, and markedly elevated usually in extra-adrenal tumours, such as carcinoma of the lung producing Cushing's syndrome.

FURTHER READING

Baxter, J. D., Rousseau, G. G. (eds) (1979) *Glucocorticoid Hormone Action. Monographs on Endocrinology*, vol. 12. New York: Springer-Verlag.

Bia, M. J., Tyler, K., De Fronzo, R. A. (1982) The effect of dexamethasone on renal electrolyte excretion in the adrenalectomised rat. *Endocrinology*, **111**, 882.

Brubaker, P. L., Baird, A. C., Bennett, H. P. J., Browne, C. A., Solomon, S. (1982) Corticotropic peptides in the human fetal pituitary. *Endocrinology*, **111**, 1150.

Fregly, M. J. (ed.) (1978) Angiotensin-induced thirst: Peripheral and central mechanisms (symposium). *Federation Proceedings*, **37**, 2667.

Givens, J. R. (1976) Hirsutism and hyperandrogenism. *Advances in Internal Medicine*, **21**, 221.

Gower, D. B. (1979) *Steroid Hormones*. London: Croom Helm.

Hornsby, P. J. (1982) Regulation of 21-hydroxylase activity of steroids in cultured bovine-adrenocortical cells: possible significance for adrenocortical androgen synthesis. *Endocrinology*, **111**, 1092.

James, V. H. T., Serio, M., Giusti, G., Martini, L. (eds) (1978) *The Endocrine Function of the Human Adrenal Cortex*. London: Academic Press.

Krieger, D. T. (1977) Regulation of circadian periodicity of plasma ACTH levels. *Annals of the New York Academy of Sciences*, **297**, 561.

Krieger, D. T., Ganong, W. F. (eds) (1977) ACTH and related peptides. *Annals of the New York Academy of Sciences*, **297**.

Peart, W. S. (1976) The renin-angiotensin system. In Parsons, J. A. (ed.) *Peptide Hormones*, p. 179–196. Baltimore: University Park Press.

Morris, D. J., Davis, R. P. (1974) Aldosterone: current concepts. *Metabolism*, **23**, 473–494.

Reid, I. A., Ganong, W. F. (1974) The hormonal control of sodium excretion. In McCann, S. M. (ed.) *Endocrine Physiology*, vol. 5, p. 205–237. Baltimore: University Park Press.

Reid, I. A., Morris, B. J., Ganong, W. F. (1978) The renin-angiotensin system. *Annual Review of Physiology*, **40**, 377.

Thompson, E. B., Lippman, M. E. (1974) Mechanism of action of glucocorticoids. *Metabolism*, **23**, 159–202.

Multiple choice questions

1. ACTH:
1. is a polypeptide which structurally resembles β MSH: both share a common core of 7 amino acids;
2. acts via cyclic AMP system, which activates (a) synthesis of m-RNA and hydroxylases, and (b) oxidation of glucose by hexose monophosphate shunt with generation of $NADPH_2$;
3. has MSH-like activity;
4. acts in steroidogenesis in the adrenal cortex in the initial step cholesterol → pregnenolone;
5. acts in concert with FSH in steroidogenesis in the testis in the initial step cholesterol → pregnenolone;
6. instilled into the appropriate area in the hypothalamus decreases production of CRF;
7. release is decreased by aldosterone;
8. release is stimulated by angiotensin II, but not by angiotensin I;
9. releasing factor also apparently increases release of β LPH;
10. stimulates the secretion of cortisol, dehydroepiandrosterone (17 KS) and aldosterone;
11. production is increased in Addison's disease and that of β LPH is unchanged.

2. With regard to the function of the pituitary-adrenocortical axis:
1. the normal secretion rate of cortisol is 50 to 75 mg/day;
2. aldosterone excretion rate provides a useful index of mineralocorticoid activity;
3. in normal subjects, 1 mg dexamethasone given between 11 p.m. and midnight will result in the 8 a.m. plasma cortisol on the following morning being suppressed to <5 μg/100 ml;
4. ^{125}I-cholesterol is a useful tracer for isotope scanning in the investigation of suspected adrenocortical tumours;
5. urinary 17-ketosteroid excretion >50 mg/day suggests the presence of an adrenocortical carcinoma in a patient with Cushing's syndrome;
6. plasma ACTH levels are suppressed in Cushing's disease with bilateral adrenocortical hyperplasia.

3. In man, the administration of aldosterone results in:
1. increased renal excretion of sodium;
2. decreased renal excretion of potassium;
3. increased renal reabsorption of sodium;
4. increased renal reabsorption of chloride;
5. increased renal reabsorption of bicarbonate;
6. expansion of extracellular volume;
7. decreased osmolarity of extracellular fluid;
8. decreased glomerular filtration rate;
9. decreased colloid osmotic pressure of plasma.

4. With regard to the renin-angiotensin system:
1. primary hyperaldosteronism is characterised by the findings of high serum sodium and potassium concentrations;
2. suppressed plasma renin activity unresponsive to salt restriction is a feature of primary hyperaldosteronism;
3. oral contraceptive administration is followed by a fall in the levels of renin substrate;
4. administration of 100 mEq/day of sodium will cause a tendency for serum potassium levels to fall in patients with primary hyperaldosteronism;
5. in the correction of hypokalaemia resulting from excessive aldosterone secretion, sodium restriction is an important measure;
6. in more than 50% of cases adrenocortical adenomata causing hyperaldosteronism are bilateral.

5. Cortisol:
1. increases glucose production by the liver;
2. increases gluconeogenesis by the liver;
3. decreases glucose utilisation by muscle cells;
4. increases free fatty acid mobilisation;
5. increases lymphocyte count in blood;
6. decreases size of the thymus;
7. concentration decreases in blood during haemorrhage;
8. increases ACTH secretion via a feedback control mechanism.

6. In the diagnosis of adrenocortical insufficiency:
1. plasma ACTH measurement will allow differentiation of tuberculous from idiopathic adrenal hypofunction;
2. serum electrolytes show characteristic changes (low Na, high K) in >50% of cases;
3. assessment of the response to exogenous ACTH is the most important diagnostic test;
4. 21-hydroxylase deficiency is the common cause of childhood cortisol deficiency;
5. ACTH levels are elevated in salt-losing congenital adrenal hyperplasia;
6. idiopathic Addison's disease has a significant association with auto-immune disorders such as pernicious anaemia.

7. Aldosterone:
1. reacts with a receptor in the cytoplasm, the complex activates the synthesis of DNA-dependent m-RNA, and this stimulates production by the ribosomes of enzyme(s) which increase the transport of Na^+ and K^+;
2. initiates the process whereby the Na^+-K^+ pump is activated in cells, e.g. in salivary and sweat glands;

3. can increase the volume and osmolarity of the ECF, however renal compensatory mechanisms limit Na^+accumulation, so oedema in hyperaldosteronism is limited;
4. can increase the osmolarity of ECF, which stimulates cells in the drinking centre and supraoptic nucleus and, as a consequence, the volume of the ECF increases;
5. production in hypovolaemia is increased (chiefly) in response to activation of the JGA-renin-angiotensin system by the reduced renal blood flow;
6. secretion can be increased by sympathetic stimulation activating the JGA-renin-angiotensin system;
7. causes Na^+ loss and K^+ retention in sweat and saliva;
8. production is increased when the $[K^+]$ in the plasma falls to 3 mEq/l;
9. production is increased when the $[Na^+]$ in distal tubular fluid in vicinity of macula densa is increased;
10. can induce tetany.

8. In patients who have received long-term, high dose corticosteroid therapy:
1. recovery of pituitary-adrenocortical function is usually complete within 4 months of ceasing the therapy;
2. after ceasing therapy pituitary ACTH secretion recovers before adrenocortical responsiveness to ACTH;
3. pituitary adrenocortical suppression does not usually occur if the dose of prednisolone is <7.5 mg daily;
4. ACTH administration during steroid therapy prevents steroid withdrawal symptoms on discontinuing the high dose steroids;
5. alternate day steroid therapy is of no benefit in preventing pituitary adrenocortical suppression;
6. responsiveness to exogenous ACTH at the conclusion of steroid therapy provides a good index of the degree of pituitary-adrenocortical suppression.

9. In the adrenal cortex:
1. of both sexes, dehydroepiandrosterone is a major product, and small amounts of testosterone and oestradiol are also formed;
2. inadequate conversion of pregnenolone → progesterone results in adrenocortical insufficiency and adrenogenital syndrome;
3. inadequate conversion of 11-deoxycorticosterone to corticosterone results in hypertension, salt retention and adrenogenital syndrome;

4. the metabolic pathways for the synthesis of progesterone, testosterone and oestradiol closely resemble those for the synthesis of these hormones in the ovary and testis;
5. the 17-ketosteroid dehydroepiandrosterone is formed and converted into other 17-ketosteroids, e.g. androsterone and epiandrosterone;
6. during pregnancy in women the secretion of progesterone is increased at least 5-fold;
7. the zona fasciculata cells respond to circulating angiotensin II by producing aldosterone;
8. after hypophysectomy, little change occurs in the zona glomerulosa and in the secretion of aldosterone;
9. the secretion of aldosterone is sensitive to changes in $[K^+]$ and in $[Na^+]$ of the plasma.

10. With respect to adrenocortical function:
1. the response to cortisol replacement in Addisonian crisis takes some hours;
2. serum calcium levels may be raised in adrenal insufficiency;
3. serum calcium levels are elevated in Cushing's disease with osteoporosis;
4. bilateral adrenal hyperplasia is the commonest cause of Cushing's syndrome;
5. no single test of adrenocortical function can with certainty be used to diagnose Cushing's syndrome;
6. serum ACTH estimation is the best method of diagnosis of Addison's disease;
7. growth 'catch-up' occurs after correction of Cushing's syndrome in children whose growth has been stunted.

11. With respect to hormones:
1. the major precursor of urinary 17KS in both sexes is dehydroepiandrosterone from the adrenal cortex;
2. in Addison's disease, both ACTH and β LPH are increased in the plasma, and the pigmentation is due chiefly to ACTH;
3. in Cushing's syndrome due to an adrenocortical tumour, the secretion of both ACTH and β LPH is reduced;
4. in primary hyperaldosteronism, hypertension and hypokalaemic alkalosis are present;
5. CRF releases ACTH and β LPH;
6. angiotensin I is converted by enzymes in the lungs and peripheral tissues to angiotensin II;
7. in primary hyperaldosteronism, hypokalaemia decreases the secretion of insulin;

8. cortisol reacts with a cytoplasmic receptor and the complex enters the nucleus and, as a consequence, m-RNA synthesis is increased and protein synthesis by the polyribosomes becomes increased.

12. In primary hyperaldosteronism due to unilateral adenoma:
1. plasma renin activity is low in the contralateral kidney venous effluent;
2. blood pressure is lowered by spironolactone;
3. aldosterone secretion is suppressed by dexamethasone;
4. aldosterone secretion is not suppressed by DOC;
5. serum calcium is lowered;
6. the polyuria is unresponsive to ADH;
7. it is important to measure aldosterone levels after potassium repletion;
8. hypokalaemia is best sought after salt administration.

13. Which of the following are true statements?
1. in the diagnosis of Cushing's syndrome the dexamethasone suppression test utilises the negative feedback of blood glucocorticoid levels at the hypothalamus and anterior pituitary;
2. in Cushing's syndrome due to ectopic production of ACTH by a tumour, suppression occurs with high but not with low dose dexamethasone;
3. in Cushing's syndrome due to an adrenal adenoma, suppression does not occur with either low or high dose dexamethasone;
4. in the metyrapone test of pituitary ACTH reserve, the drug is an inhibitor of 11 β-hydroxylation in the adrenal cortex;
5. in the metyrapone test of pituitary ACTH reserve, a normal test is associated with a 50% fall in urinary 17-hydroxycorticosteroid metabolites on the day of administration of the drug.

14. Cortisol differs from growth hormone in that it:
1. increases excretion of a water load in adrenocortical insufficiency;
2. is produced during acute haemorrhage (e.g. 15% of circulating blood volume);
3. can cause hyperglycaemia and ketonaemia;
4. is produced in increased amounts during surgical stress;
5. increases the transport of amino acids across some cell membranes;
6. increases hepatic glucogenesis.

15. With respect to adrenal medulla hormones:
1. catecholamines mobilise energy substrates, chiefly glucose (liver), FFA (fat cells and muscle), and ketone bodies (liver), and these effects are increased in exercise;
2. the biosynthetic pathway in hormone synthesis is: tyrosine → DOPA → dopamine → noradrenaline → adrenaline;
3. the enzyme for the conversion of noradrenaline → adrenaline, is present virtually only in the adrenal medulla hence, adrenaline synthesis occurs only in the adrenal medulla and not postganglionic adrenergic nerve endings;
4. the catecholamines are stored as granules in vesicles complexed to a specific protein and ATP; any free hormone is inactivated by monoamine oxidase (MAO) in the mitochondria;
5. on vascular smooth muscle α receptors initiate constriction and β receptors, dilatation;
6. both α and β receptors respond to both hormones: noradrenaline stimulates chiefly α receptors, and adrenaline stimulates both about equally;
7. glucocorticoids from adrenal cortex induce formation of the enzyme for the conversion of noradrenaline to adrenaline and hence highest concentrations of adrenaline are in outer medulla.

16. Comparing the actions of adrenaline and noradrenaline:
1. adrenaline is more effective in constricting arterioles in skin/mucous membranes;
2. adrenaline is more effective in increasing cardiac output and pulse pressure and reducing peripheral resistance;
3. adrenaline is a more effective bronchodilator;
4. noradrenaline does not increase hepatic glycogenolysis;
5. adrenaline dilates vessels in skeletal muscle;
6. adrenaline does not mobilise FFA from fat cells;
7. both decrease insulin secretion;
8. adrenaline is more effective in raising the plasma glucose level.

17. The following statements apply to patients suffering from phaeochromocytoma:
1. neurofibromatosis is associated in about 5% of patients;
2. calcitonin measurement may be a relevant investigation;
3. postural hypotension is not an uncommon symptom;
4. the free thyroxine index is a useful test in differential diagnosis from thyrotoxicosis;
5. the initial screening test for the disease is to administer phentolamine (1 mg) and observe the blood pressure response;

6. blood volume replacement may be necessary pre-operatively;
7. glucagon administration produces an increase in catecholamine levels.

18. With respect to adrenal medulla hormones:
1. sympathetic (splanchnic) stimulation releases Ach at the preganglionic ending contiguous to the phaeochromocyte, Ca^{++} influx occurs, and a quantum of hormones is released from the vesicles into the adrenal venous blood;
2. circulating hormones are metabolised by catechol-O-methyl transferase (COMT) which is present in many tissues;
3. the chromaffin cells which synthesise adrenaline are different from those which produce noradrenaline, and are innervated by different preganglionic sympathetic cholinergic fibres;
4. adrenaline/noradrenaline secretion ratio normally is about 4:1 (in man);
5. the phaeochromocytes respond selectively to stimulation: e.g. hypotension reflexly releases noradrenaline (mainly); hypoglycaemia and exposure to cold reflexly release adrenaline (mainly); and the response to hypothalamic stimulation depends on the area excited;
6. adrenaline secretion is increased in hypoglycaemia: this helps to maintain homeostasis of the plasma glucose level by decreasing insulin secretion, increasing glycogenolysis in liver, and increasing gluconeogenesis from lactate in liver.

Answers

1. 1, 2, 3, 4, 6, 9, 10
2. 2, 3, 4, 5
3. 3, 4, 5, 6, 9
4. 2, 4, 5
5. 1, 2, 3, 4, 6
6. 3, 4, 5, 6
7. 1, 2, 3, 4, 5, 6, 10
8. 2, 3
9. 1, 2, 3, 4, 8, 9
10. 1, 2, 4, 5
11. 1, 2, 3, 4, 5, 6, 7, 8
12. 1, 2, 4, 6, 7, 8
13. 1, 3, 4
14. 1
15. 1, 2, 3, 4, 5, 6, 7
16. 2, 3, 5, 7, 8
17. 1, 2, 3, 4, 6, 7
18. 1, 2, 3, 4, 5, 6

5. Testis

FUNCTIONS OF TESTIS

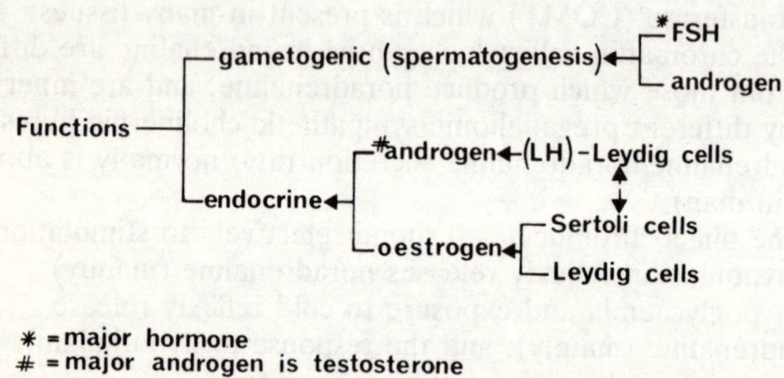

SPERMATOGENESIS

74 days from spermatogonium to sperm cell; spermatogonia undergo a number of mitotic divisions and then mature into next stage of development.

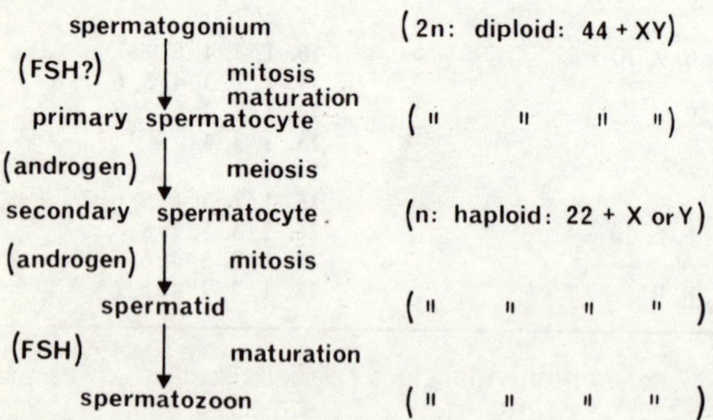

Fig. 5.1 Spermatogenesis

Developing sperm cells are nourished by glycogen-charged Sertoli cells.

Testosterone is necessary for complete maturation of sperm cells.

Function of Leydig cells and Sertoli cells in spermatogenesis
FSH and androgen are involved in spermatogenesis, but there is no convincing evidence that: (1) FSH has *direct* effect on developing germ cells and (2) androgen has *direct* effect on germ cell maturation. The effects of both hormones appear to be mediated indirectly.

$$\text{LH} \xrightarrow[\text{Leydig cell}]{\text{LH receptor}} \text{testosterone}$$

Testosterone is present in high concentration at periphery (basal aspect) of seminiferous tubules (germinal epithelium) and exerts an essential paracrine function on spermatogenesis (gametogenesis).

FSH —(FSH receptor / Sertoli cell)→ unidentified ? substance →
- ↑ testosterone by Leydig cells
- ↑ sensitivity of Leydig cells to LH

Sertoli cells: (1) form tight junctions with adjacent Sertoli and other cells → blood/lymph barrier at basal aspect of tubule/duct system → protective seal which prevents back-diffusion of sperm autoantigens from lumen of tubules into blood stream. This Sertoli-Sertoli barrier normally prevents immune response and, thus, formation of sperm autoantibodies; and (2) sustain spermatogenesis.

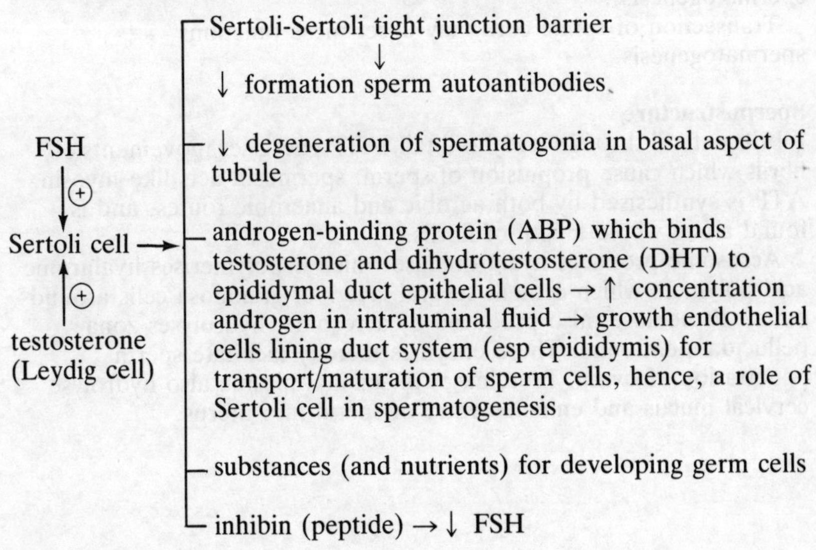

Sertoli cell secretes Müllerian regression factor in embryonic life (see p. 233).

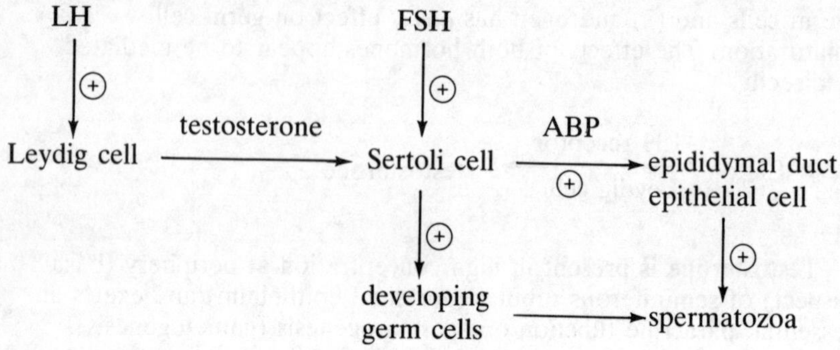

Effects of temperature

Spermatogenesis is sensitive to temperature change and is inhibited by increased temperature (e.g. in cryptorchidism, spermatogenesis is suppressed, but testosterone secretion may be only slightly reduced). Scrotal reflexes cause muscles of scrotum to contract or relax in response to environmental temperature. Scrotum is copiously supplied with sweat glands and has a thermoregulatory function.

Countercurrent blood flow in spermatic arteries and pampiniform plexus of spermatic veins in scrotum → countercurrent exchange of heat and testosterone → ↓ temperature (improves efficiency of cooling) and ↑ testosterone concentration in testis (↑ spermatogenesis).

Transection of spinal cord may upset these functions → ↓ spermatogenesis.

Sperm structure

Fibrils extend through body and tail. ATP energises movements of fibrils which cause propulsion of sperm: spermosin acts like myosin. ATP is synthesised by both aerobic and anaerobic routes, and is found along whole length of fibrils.

Acrosome produces hyaluronidase which depolymerises hyaluronic acid polymers which cement several layers of granulosa cells around ovum. Proteinase, also produced by acrosome, hydrolyses zona pellucida membrane. These enzymes thereby facilitate sperm penetration of ovum. Proteinases in seminal plasma also hydrolyse cervical mucus and enhance entry of sperm into uterus.

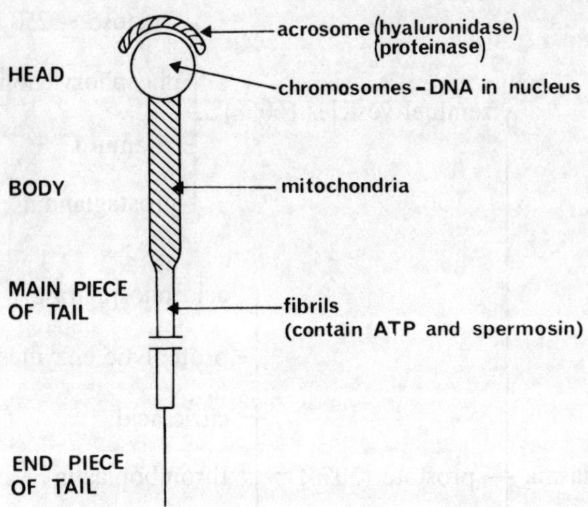

Fig. 5.2 Sperm structure

SEMEN

Seminal plasma provides *protective* medium and *nutritive* source for sperm after ejaculation. Since semen consists of sperm cells and seminal plasma it is not stored as such in the body but is normally produced on ejaculation.

Composition and formation

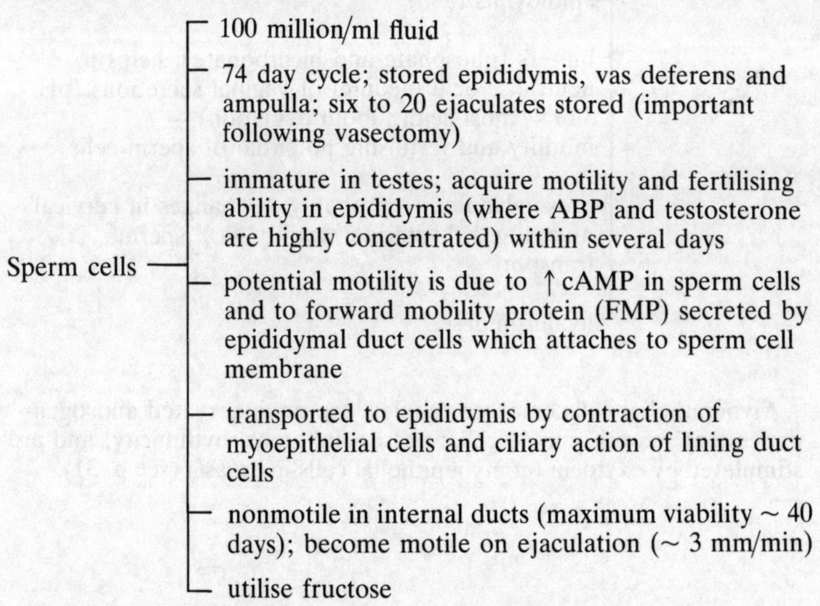

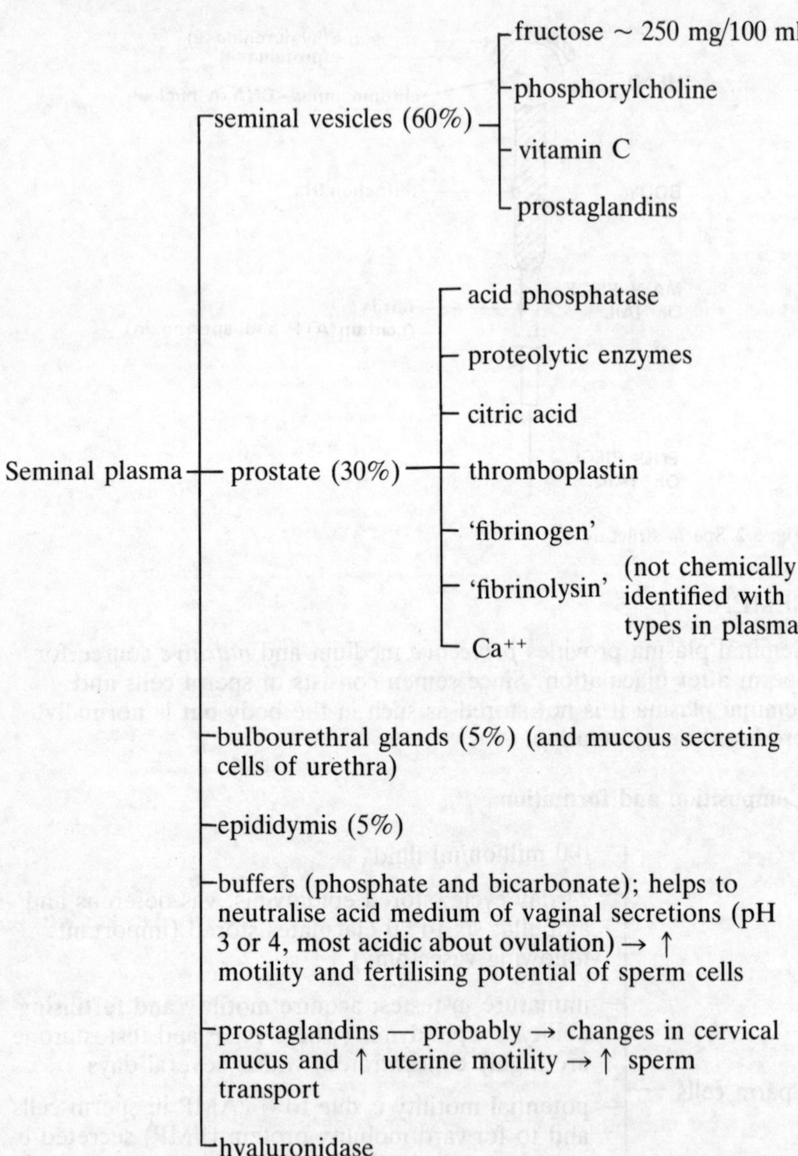

Myoepithelial cells embrace tubules; are noninnervated androgen-dependent smooth muscle cells; possess inherent rhythmicity; and are stimulated by oxytocin (cf myoepithelial cells in breast) (see p. 31).

FERTILISATION

See Figure 5.3.

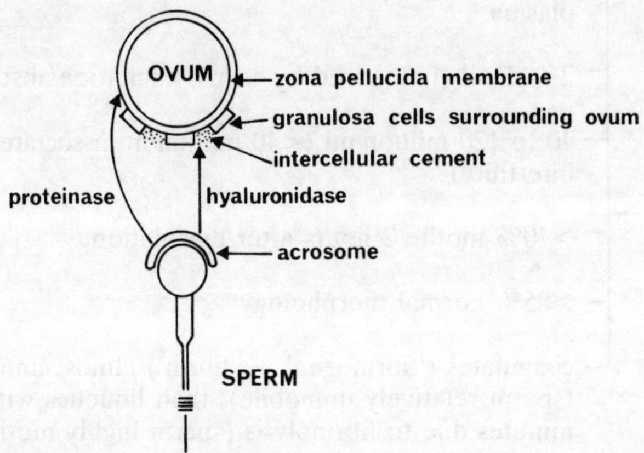

Fig. 5.3 Fertilisation

Species specificity of fertilisation is due to a receptor on ovum membrane which reacts with a species-specific protein (bindin) on homologous sperm cell membrane. Sperm bindin-receptor interaction stimulates the ovum by releasing it from inhibition, and changes occur in its membrane permeability, intracellular fluid/electrolyte composition and pH.

It is hyaluronidase in acrosome of the individual sperm cell which is important in sperm penetrating between corona cells and not total amount of hyaluronidase present in the semen.

Secretions from accessory structures increased by parasympathetic stimulation during sexual excitation.

Substances in seminal plasma and in vagina increase fertilising power of sperm by removing inhibitory material from the acrosome (i.e. process of capacitation). Sperm are viable for several hours in vagina and 24 to 48 hours in uterus and Fallopian tubes. Unidentified substances in vagina may cause rapid death of sperm; and sperm immobilising antibodies may be present in female reproductive tract which account for some unexplained cases of infertility.

Fertilisation occurs usually in Fallopian tube and probably only 10 to 20 sperm cells reach vicinity of ovum. It takes ~ 4 days for fertilised ovum to be propelled by ciliary movements and peristalsis to uterus and ~ 3 days for implantation to be effected: nidation occurs at height of development of corpus luteum and of endometrium.

EJACULATE

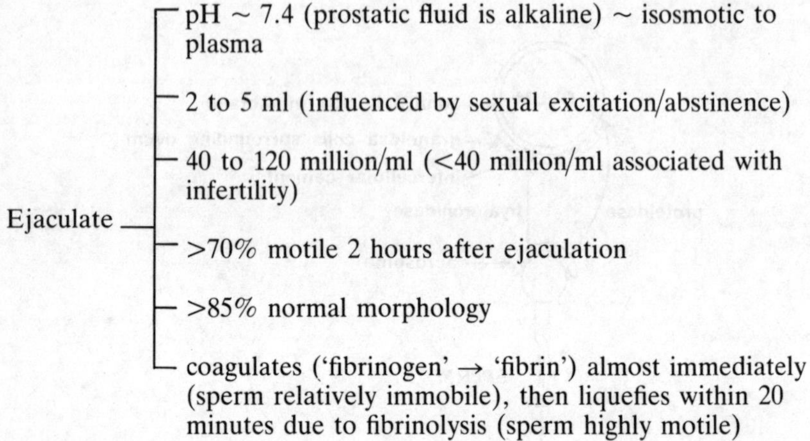

VASECTOMY

After vasectomy, sperm continue to be formed but are reabsorbed (? phagocytosed) in epididymis. This may lead to formation of sperm autoantibodies which cause infertility in most cases where re-anastomosis is carried out. No loss of libido or potentia occurs after vasectomy, and long term effects on endocrine function are still not clear — possibly plasma LH rises slowly and testosterone falls due to damage to Leydig cell function. Six to 20 ejaculations after vasectomy are required to remove stored sperm: fertility may thus be possible for several months following operation.

TESTOSTERONE

Mechanism of action

Testosterone is converted to dihydrotestosterone in the cytoplasm which then reacts with a cytoplasmic receptor; this complex enters the nucleus and initiates increased synthesis of m-RNA, leading to increased protein production by the ribosomes.

Effects of testosterone

1. Development of external genitalia (penis, scrotum); development — secretion internal sex organs (seminal vesicles, prostate, etc.).
2. Development of secondary sex characteristics (hair, voice, muscle, etc.), and psychosexual behaviour.
3. Descent of testes into scrotum.

4. Increases libido and facilitates the neurogenic mechanisms responsible for erection and ejaculation (sexual performance).
5. Induces masculinisation (in fetus) of external genitalia, and masculinisation of hypothalamus (in late fetal-neonatal period) with respect to subsequent psychosexual behaviour patterns and gonadotrophin release during reproductive life (concentration of testosterone in fetal testes peaks at time of sexual differentiation of external genitalia and concentration declines soon after maternal HCG plasma level declines).
6. Increases sebaceous gland secretion (→ acne); NaCl and H_2O reabsorption (kidneys); protein synthesis; bone protein matrix and Ca deposition; and BMR (5 to 10 %).
7. Stimulates growth (pubertal growth spurt in boys).
8. Enhances maturation of bone and eventually causes closure of epiphyses.
9. Reduces level of thyroxine binding proteins and so reduces level of total thyroxine and PBI. The T3-resin uptake is increased (already mentioned under *Thyroid*, p. 54).

Secretion of testosterone

Testosterone is secreted by the Leydig cells, and is postulated also to be produced directly in seminiferous tubules and pregnenolone sulphate may serve as its major source.

Age and testosterone

At age 30 a slow decline in plasma testosterone commences, and accelerates beyond age 55. Nevertheless, the normal ranges at ages 30 and 60 show considerable overlap. Rising levels of LH in the older age groups tend to maintain testosterone production. FSH also rises with age.

CONTROL OF TESTIS FUNCTION

This is shown in Figure 5.4.

It has been postulated that a second negative feedback loop between testis and hypothalamus/anterior pituitary exists via an unidentified substance (inhibin) secreted by the seminiferous tubules which selectively inhibits FSH secretion. However, there is no direct evidence for independent control of FSH and LH secretion.

LH response is mediated by cAMP mechanism. LH (or HCG) receptor on plasma membrane of Leydig cell has high binding affinity for the hormone and can be used in a radioligand assay.

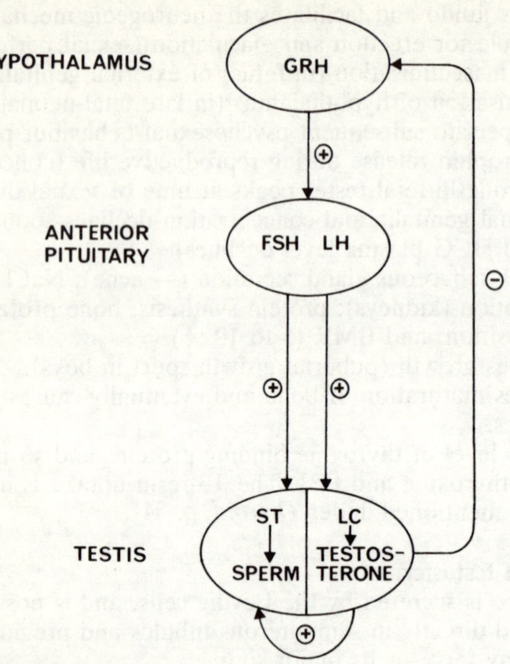

Fig. 5.4 Control of testis function

Key: GRH = gonadotrophin releasing hormone
ST = seminiferous tubules
LC = Leydig cell
FSH stimulates spermatogenesis; effect enhanced by testosterone.
LH stimulates testosterone secretion (after priming effect of FSH on Leydig cells).
Castration (adult) increases secretion of FSH and LH

Transport and metabolism of testosterone

Secretion rate — 5 to 9 mg/day (male, adult)

Plasma levels
- 3 to 12 ng/ml plasma (male, adult)
- 0.2 to 0.7 ng/ml plasma (female, adult)

Transported
- bound to *TeBG (85%) (i.e. testosterone-oestradiol binding globulin) which is made in liver
- bound to albumin (14%)
- free (1%)

* = also called steroid hormone-binding globulin (SHBG)

Testosterone is converted into dihydrotestosterone (DHT) in androgen-dependent tissues (and as such is more active). In liver testosterone is converted into 17-oxosteroids (relatively nonandrogenic steroids e.g. androsterone, epiandrosterone, aetiocholanolone), which are conjugated and excreted in the urine as glucuronides and sulphates.

DHT is twice as potent as testosterone; has a higher affinity for SHBG; is in low concentration in blood; and plays a role in development of male generative tract (see p. 232).

Synthetic androgens include testosterone propionate and methyl testosterone; anabolic androgens (19-nor steroids: lack methyl group in 19 position), e.g. norethandrolone or Nilevar, have high anabolic: androgenic ratios; and antiandrogens, e.g. cyproterone, block androgen effect by binding to androgen receptor.

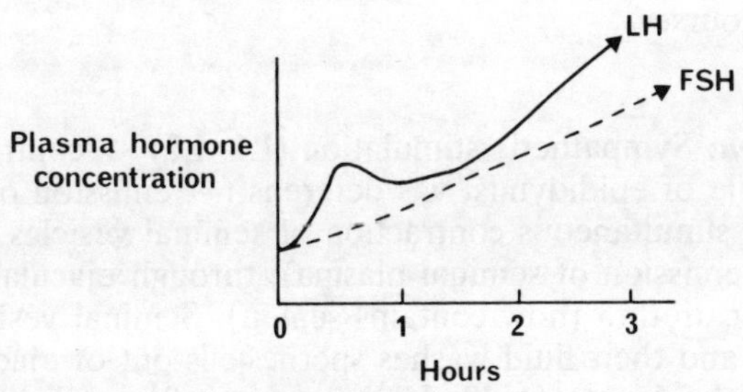

Fig. 5.5 GRH and gonadotrophin release

GRH causes release of FSH from pituitary (but without early peak release as with LH in infusion studies — suggesting GRH releases stored LH but not stored FSH).

A circadian variation in plasma testosterone (highest 6 to 7 a.m.; lowest 7 to 8 p.m.) is present, but there is no evidence for a circadian rhythm of plasma LH.

During REM sleep, penile erections occur and testosterone secretion increases, but the pulsatile pattern of FSH/LH secretion which is found during the day is unaltered.

In blind men, plasma testosterone and FSH/LH levels are similar to those of normal men and, therefore, are not light-dependent.

PHYSIOLOGY OF COITUS

Male sexual response

Erection

Parasympathetic excitation → dilatation of penile arteries → ↑ blood flow into venous sinusoids of corpora cavernosa and corpus spongiosum. Compression of dorsal vein and ↑ tone in bulbospongiosus and ischiocavernosus muscles → ↓ blood flow out of erectile tissue.

Sympathetic stimulation → vasoconstriction of penile arteries → deturgesence.

Lubrication

Parasympathetic excitation → mucus secretion (urethral glands and bulbourethral glands); of minor importance in vaginal lubrication during intercourse.

Ejaculation

a. *Emission*. Sympathetic stimulation (L1, L2) → contraction of smooth muscle of epididymis, vas deferens (→ emission of sperm), and (almost) simultaneous contraction of seminal vesicles and prostate (→ emission of seminal plasma), through ejaculatory duct into posterior urethra (now contains semen). Seminal vesicles contract last and their fluid washes sperm cells out of ejaculatory ducts. Internal sphincter of bladder constricts. Semen may ooze from penis during sympathetic stimulation (pudendal) at operation; and during intercourse occasionally weak emission of semen from penis can occur without strong somatically-induced pulsatile ejaculation (due to urethral peristalsis).

b. *Ejaculation proper*. Somatic (pudendal-superficial perineal branches) (S1, S2) stimulation → five to 15 rhythmic contractions of bulbospongiosus and ischiocavernosus muscles at bulb/crus of penis which eject semen under considerable pressure. These muscles may also be contracted voluntarily.

The initial contractions are the most expulsive, and their time relationship is similar to those occurring in muscles in the perineal region during orgasm in the female. Other perineal/pelvic muscles contract rhythmically during ejaculation.

During intercourse contraction of the dartos muscles and increased blood flow into the scrotum causes the integument to thicken; contraction of the cremaster muscles elevates the testes which (increased in size due to vasocongestion) become less mobile in a smaller scrotal sac.

CONTROL OF MALE SEXUAL RESPONSE

This is illustrated in Figure 5.6.

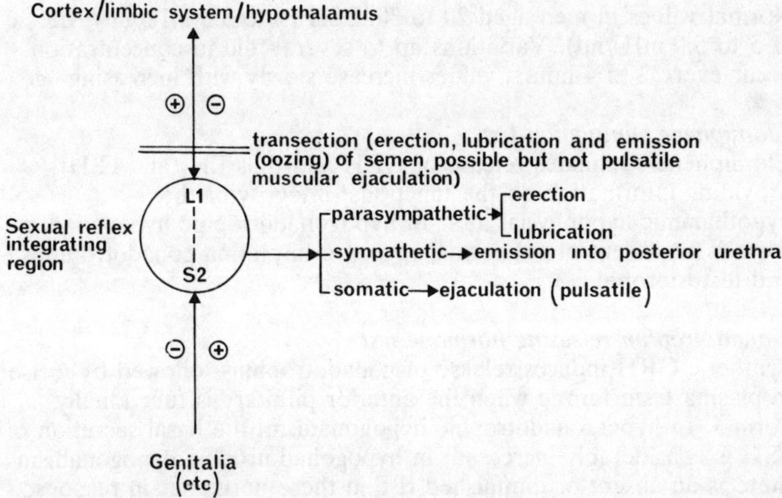

Fig. 5.6 Control of male sexual response

HYPOGONADISM

Male hypogonadism may be either testicular in origin (primary) with high plasma LH and FSH levels, or due to failure of pituitary secretion of LH and FSH (secondary). For restoration of male sexual function, strength, vigour and hairgrowth, synthetic androgen is given (e.g. methyltestosterone orally). In the treatment of infertility, human FSH derived from post-menopausal female urine may be used in secondary hypogonadism. In some cases, clomiphene citrate is successful in stimulating endogenous FSH and LH secretion. Mesterolone, a weak synthetic androgen, raises fructose levels in seminal plasma (without the complete suppression of LH and FSH secretion which is produced by testosterone) and is sometimes effective in oligospermia (10 to 20 million/ml).

TESTICULAR FUNCTION TESTS (PRINCIPLES ONLY)

Gonadotrophins

Radioimmunoassay of plasma gonadotrophins
Normal values in men aged 20 to 30: FSH (1 to 3.5 mIU/ml), LH (1.5 to 5.0 mIU/ml). Variations up to several-fold in concentration occur every 2 to 4 hours; values increase slowly with increasing age.

Clomiphene stimulation test
Clomiphene stimulates release of GRH and thus LH (and FSH), providing information on the functional integrity of the hypothalamic-hypophysial axis. In hypogonadotrophic hypogonadism, there is an absent or subnormal response in plasma gonadotrophins and testosterone.

Gonadotrophin releasing hormone test
Synthetic GRH induces release of gonadotrophins followed by a rise in plasma testosterone when the anterior pituitary is functionally normal. In hypergonadotrophic hypogonadism, the basal secretion of FSH is considerably increased; in hypogonadotrophic hypogonadism there is an absent or diminished rise in these hormones in response to GRH.

In primary and secondary hypogonadism plasma testosterone is low. In primary hypogonadism plasma gonadotrophin levels are elevated, whereas in secondary hypogonadism they are depressed.

Testicular hormones

Measurement of plasma testosterone
Radioimmunoassay and competitive protein-binding assay measurements of plasma testosterone give values of 4 to 10 ng/ml; 12 to 30 nmoles/l (men aged 20 to 50) with moment to moment fluctuations.

HCG stimulation test
HCG is administered and plasma testosterone levels are determined. May help to differentiate between primary hypergonadotrophic hypogonadism and secondary hypogonadotrophic hypogonadism; and especially useful in cryptorchidism where testosterone secretion is induced when testicular tissue is present.

Measurement of plasma oestradiol
Values of 5 to 30 pg/ml; < 100 pmol/l (men aged 20 to 30).

Measurement of urinary oestrogens
Values of 8 to 15 μg/24 hours (men aged 20 to 30).

Analysis of semen

Other investigations include testicular biopsy and sperm antibodies.

FURTHER READING

Acott, T. S., Hoskins, D. D. (1978) Bovine sperm forward motility protein. *Journal of Biological Chemistry*, **253**, 6744–6750.

Bardin, C. W. (1978) Pituitary-testicular axis. In Yen, S. S. C., Jaffe, R. B. (eds) *Reproductive Endocrinology*. Philadelphia: Saunders.

Bronson, F. H., Desjardins, C. (1982) Endocrine responses to sexual arousal in male mice. *Endrocinology*, **111**, 1286.

Burger, H. G., de Kretser, D. M., Hudson, B., Wilson, J. D. (1981) Effects of preceding androgen therapy on testicular response to human pituitary gonadotropin in hypogonadotropic hypogonadism: A study of three patients. *Fertility and Sterility*, **35**, 64.

Danzo, B. J., Taylor Jr, C. A., Eller, B. C. (1982) Some physicochemical characteristics of photoaffinity-labelled rabbit androgen-binding protein. *Endrocinology*, **111**, 1270.

Davies, T. F., Dufau, M. L., Catt, K. J. (1978) Gonadotrophin receptors: Characteristics and clinical applications. *Clinical Journal of Obstetrics and Gynecology*, **5**, 329–362.

de Jong, F. H. (1979) Inhibin — fact or artifact. *Molecular and Cellular Endocrinology*, **13**, 1–10.

Dufau, M. L., Catt, K. J. (1978) Gonadotropin receptors and regulation of steroidogenesis in the testis and ovary. *Vitamins and Hormones*, **36**, 461–592.

Ewing, L. L., Robaire, B. (1978) Endogenous antispermatogenic agents: prospects for male contraception. *Annual Review of Pharmacology and Toxicology*, **18**, 167.

Fawcett, D. W. (1975) Ultrastructure and function of the Sertoli cell. In Hamilton, D. W., Greep, R. O. (eds) *Handbook of Physiology*, section 7: Endocrinology, vol. V: Male Reproductive System, p. 21–56. Baltimore: Williams & Wilkins.

Franchimont, P., Roulier, R. (1977) Gonadotropin secretion in male subjects. In Martini, L., Besser, G. M. (eds) *Clinical Neuroendocrinology*, p. 197–212. New York: Academic Press.

Fritz, I. B. (1978) Sites of action of androgens and follicle-stimulating hormone on cells of the seminiferous tubule. In Litwack, G. (ed.) *Biochemical Actions of Hormones*, vol. 5, p. 249–281. New York: Academic Press.

Greep, R. O., Koblinsky, M. A. (1977) *Frontiers in Reproduction and Fertility Control* (Ford Foundation). Cambridge, Mass.: MIT Press.

Gunsalus, G. L., Musto, N. A., Bardin, C. W. (1980) Bi-directional release of a Sertoli cell product, androgen binding protein, into the blood and seminiferous tubule. In Steinburger, E., Steinberger, A. (eds) *Testicular Development, Structure and Function*, p. 291–297. New York: Raven.

Huhtaniemi, I. T., Clayton, R. N., Catt, K. J. (1982) Gonadotrophin binding and Leydig cell activation in the rat testis in vivo. *Endocrinology*, **111**, 982.

Huhtaniemi, I.T., Nozu, K., Warren, D. W., Dufau, M. L., Catt, K. J. (1982) Acquisition of regulatory mechanisms for gonadotrophin receptors and steroidogenesis in the maturing rat testis. *Endocrinology*, **111**, 1711.

Lepow, I. H., Crozier, R. (eds) (1979) *Vasectomy: Immunologic and Pathophysiologic Effects in Animals and Man*. New York: Academic Press.

Nelson, L., Young, M. J., Gardner, M. E. (1980) Sperm motility and calcium transport: A neurochemically controlled process. *Life Sciences*, **26**, 1739–1749.

Mainwaring, W. I. P., Mann, T. (1976) *The Mechanism of Action of Androgens*. New York: Springer Verlag.

Means, A. R., Dedman, J. R., Tash, J. S., Tindall, D. J., van Sickle, M., Welsh, M. J. (1980) Regulation of the testis Sertoli cell by follicle stimulating hormone. *Annual Review of Physiology*, **42**, 59–70.

Mohri, H., Yanagimashi, R. (1980) Characteristics of motor apparatus in testicular epididymal and ejaculated spermatozoa. *Experimental Cell Research*, **127**, 191–196.

Moltz, K., Rommler, A., Post, K., Schwartz, V., Hammerstein, J. (1980) Medium dose cyproterone acetate (CPA): Effects on hormone secretion and on spermatogenesis in men. *Contraception*, **21**, 393–414.

Musto, N. A., Gunsalus, G. L., Bardin, C. W. (1980) Purification and characterization of androgen binding protein from the rat epididymis. *Biochemistry*, **19**, 2853–2860.

Ohno, S. (1977) Control of meiotic processes. In Troen, P., Nankin, H. R. (eds) *The Testis in Normal and Infertile Men*, p. 1–8. New York: Raven Press.

Setchell, B. P. (1978) Spermatogenesis. In Setchell, B. P. (ed.) *The Mammalian Testis*, p. 181–185. Ithaca, NY: Cornell University Press.

Vigersky, R. A., Kono, S., Sauer, M., Lipsett, M. B., Loriaux, D. L. (1979) Relative binding of testosterone and estradiol to testosterone-estradiol-binding globulin. *Journal of Clinical Endocrinology and Metabolism,* **49,** 899.

Multiple choice questions

1. With respect to the functions of the testis:

1. HCG stimulates the Leydig cells to secrete testosterone, and this occurring about the 7th week of fetal life is essential for differentiation of the male external genitalia;
2. an anencephalic male fetus develops male external genitalia because of testosterone secreted by its Leydig cells in response to HCG stimulation;
3. the secretion of testosterone by the Leydig cells shows a diurnal rhythm: lowest levels 7 to 8 p.m., highest 7 to 8 a.m.;
4. testosterone is the most androgenic 17-ketosteroid;
5. testosterone is converted to dihydrotestosterone in the cytoplasm, which complexes with a cytoplasmic receptor, this complex enters the nucleus and initiates increased synthesis of m-RNA;
6. some Sertoli cell tumours secrete oestrogen;
7. following bilateral orchidectomy (in a young adult) FSH plasma level remains unchanged, because the secretion of testosterone by the adrenal cortex is increased and this maintains homeostasis of FSH secretion;
8. following bilateral orchidectomy (in a young adult) the oestrogen plasma level remains unchanged, because the secretion of oestrogen by the adrenal cortex is increased and this maintains homeostasis of oestrogen secretion;
9. following bilateral orchidectomy (in a young adult) the oestrogen plasma level falls: this is chiefly because the main source of oestrogen (in man) is derived from androgen and, to a lesser extent, is secreted by the testis.

2. Semen:

1. when ejaculated, coagulates within several minutes; then liquefies within 15 to 20 minutes, because it contains 'fibrinolysin' which is produced by the prostate;
2. contains sperm cells, whose potentia to fertilise and become motile increases during their transit through the epididymis;
3. is not stored in the seminal vesicles, epididymis or vas deferens;
4. contains sperm cells, whose transit to the epididymis is due apparently to ciliary movement and peristalsis of the conducting tubules;
5. has little buffering capacity, and hence sperm cells are rapidly destroyed in the acid medium of the vagina;
6. is ejaculated in response to somatic reflexes which operate through S1-S2 cord level.

3. With regard to disorders of male reproductive function:
1. aplasia of the germinal epithelium is usually accompanied by a significant rise in the levels of serum FSH;
2. serum FSH levels are frequently normal in patients with Klinefelter's syndrome;
3. the administration of human chorionic gonadotrophin (HCG) with measurement of plasma testosterone concentration allows the differentiation of anorchia from bilateral cryptorchidism;
4. patients who are prepubertal show a significant rise in serum FSH and LH levels following clomiphene administration;
5. testosterone administration (100 μg IM weekly) to normal men will cause a significant depression in sperm count;
6. serum FSH and LH measurement are of no value in the differential diagnosis of the cause of azoospermia.

4. Erection of the penis:
1. can occur following spinal transection at T1-T12;
2. occurs in response to dilatation of penile arteries which are innervated by sympathetic cholinergic nerves;
3. subsides in response to contriction of penile arteries mediated by noradrenergic sympathetics;
4. is essential for ejaculation to occur;
5. can occur during several years after bilateral orchidectomy in the young adult;
6. is accompanied by an increase in penile temperature;
7. cannot occur in the neonate;
8. becomes impaired following vasectomy, and this is apparently the result of reduced testosterone production by the testes.

5. Ejaculation of semen:
1. is preceded by emission of sperm cells (in a small amount of fluid) from stores which are chiefly in the epididymis, vas deferens and seminal vesicles;
2. occurs as a result of rhythmic contractions of bulbospongiosus and ischiocavernosus muscles; and during orgasm in the female, similar contractions of perineal muscles occur;
3. as a strong, pulsatile event is impaired following spinal transection at T1-T12;
4. is terminated by peristaltic contractions of the penile urethra;
5. is imminent when the testes are maximally elevated by contraction of the cremaster muscles;
6. is most forceful when the testes are maximally elevated;
7. is due largely to rhythmic muscle contractions of the penile urethra and prostate.

6. A 25-year-old male patient has small testes and sparse pubic hair. Which of the following tests will be useful in making a differential diagnosis?
1. plasma testosterone;
2. serum FSH and LH;
3. LRF stimulation test;
4. urine 17-oxosteroids (ketosteroids);
5. leucocyte karyotype;
6. semen analysis.

7. Semen contains:
1. buffers which help to neutralise the acid medium of the vagina;
2. fructose;
3. prostaglandins;
4. sperm cells which have been stored in the seminal vesicles;
5. sperm cells which have a maturation cycle (spermatogonia to spermatozoa) of 14 days in man;
6. sperm cells which produce hyaluronidase and proteinase in their acrosomes.

8. Testosterone is:
1. converted in the liver to 17-ketosteroids (17-oxosteroids);
2. synthesised from cholesterol in the Leydig cells of the testis;
3. synthesised in the ovary and is the precursor of oestradiol;
4. formed in small amounts in the adrenal cortex;
5. involved in spermatogenesis;
6. a 17-ketosteroid (17-oxosteroid).

9. Male infertility:
1. may be due to a sperm count of less than 20 million/ml;
2. may be due to impotence and lack of libido;
3. usually has an endocrine basis;
4. is usually corrected by testosterone injections;
5. may be a manifestation of gonadal dysgenesis.

10. Sperm cell:
1. motility (mobility) is associated with a rise in their level of cAMP;
2. motility is enhanced by forward mobility protein (FMP) which is secreted by the Leydig cells;
3. maturation is completed in the seminiferous tubules;
4. transport from seminiferous tubules to epididymis is due to contraction of myoepithelial cells and ciliary action of lining duct cells;

5. autoantibody production normally is minimised by Sertoli-Sertoli cell barrier around seminiferous tubules;
6. fertilising ability is associated with high testosterone and androgen-binding protein (ABP) levels in epididymal duct cells.

11. During spermatogenesis (gametogenesis):
1. FSH stimulates Sertoli cells to produce androgen-binding protein (ABP);
2. Leydig cells secrete testosterone which stimulates Sertoli cells to produce ABP;
3. ABP increases concentration of testosterone in epididymal duct cells;
4. testosterone is aromatised into oestrogen;
5. testosterone is converted into dihydrotestosterone (DHT) in internal sex organs, e.g. vas deferens;
6. spermiogenesis begins with the spermatid stage and is characterised by an absence of cell division.

12. With respect to the testis:
1. spermiogenesis is that part of spermatogenesis which begins with the spermatid stage;
2. Sertoli cells secrete androgen-binding protein (ABP) and substances involved in spermatogenesis;
3. myoepithelial cells surrounding the seminiferous tubules are noninnervated, androgen-dependent, smooth muscle cells;
4. dihydrotestosterone is more active than testosterone and is bound to steroid hormone-binding globulin (SHBG);
5. role of FSH in spermatogenesis appears to be due to its effects on Sertoli cells, and not to its effects directly on the developing germ cells.

Answers

1. 1, 2, 3, 5, 6, 9
2. 1, 2, 3, 4, 6
3. 1, 3, 5
4. 1, 3, 5, 6
5. 2, 3, 4, 5, 6
6. 1, 2, 3, 5, 6
7. 1, 2, 3, 6
8. 1, 2, 3, 4, 5
9. 1, 5
10. 1, 4, 5, 6
11. 1, 2, 3, 4, 5, 6
12. 1, 2, 3, 4, 5, 6

---

6. Ovary

GONADOTROPHINS

See Chapter 1 for details.

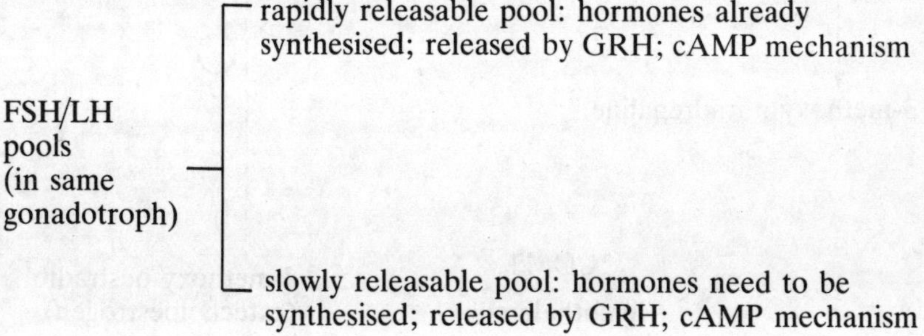

Note

- FSH and LH are secreted by same gonadotroph
- GRH stimulates release of both FSH and LH
- FSH/LH concentration ratio in blood varies during course of menstrual cycle
- FSH/LH secretion ratio varies during course of menstrual cycle: this dissociation could be due to shifting oestrogen/progesterone concentration ratio in blood altering secretory response of gonadotroph to GRH or to folliculostatin (ovarian inhibin) which inhibits FSH, and not LH, secretion (but mechanism is not clear)

CATECHOLOESTROGENS, CATECHOLAMINES AND GRH RELEASE

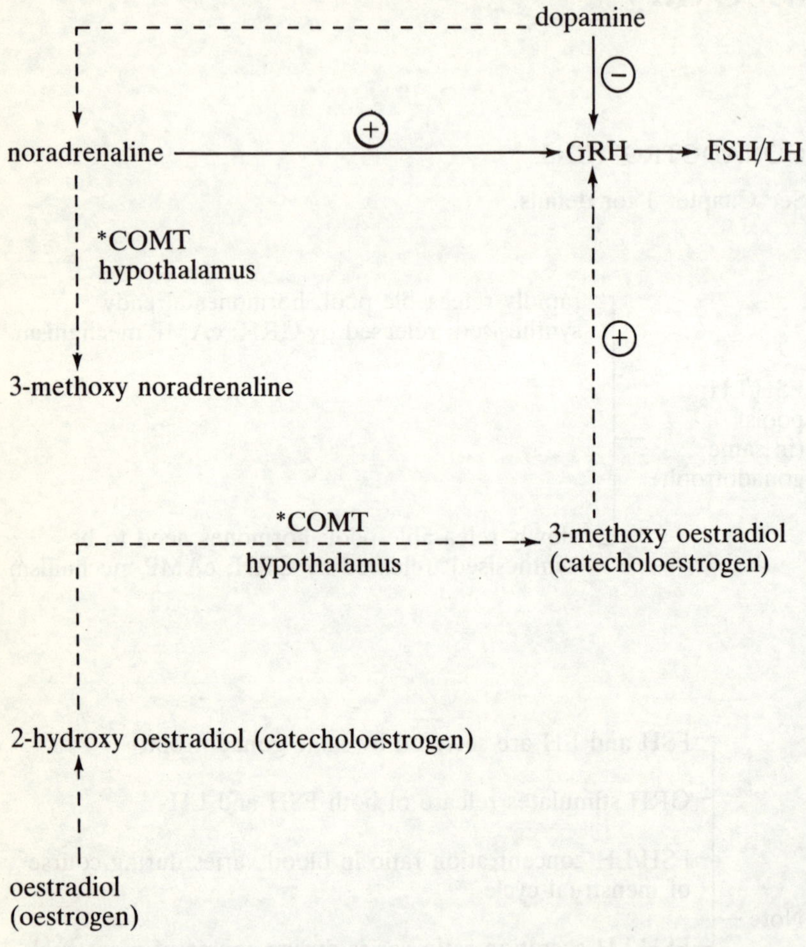

Catecholoestrogens are in high concentration in hypothalamus and pituitary; and could stimulate GRH secretion by competitively inhibiting breakdown of noradrenaline by COMT or inhibiting oestrogen inhibition of GRH.

*COMT = catechol O methyl transferase

MENSTRUAL OR REPRODUCTIVE CYCLE

By convention, the 28-day cycle is timed from the start of menstrual bleeding, and ovulation occurs on day 14.

1. Hormonal cycle

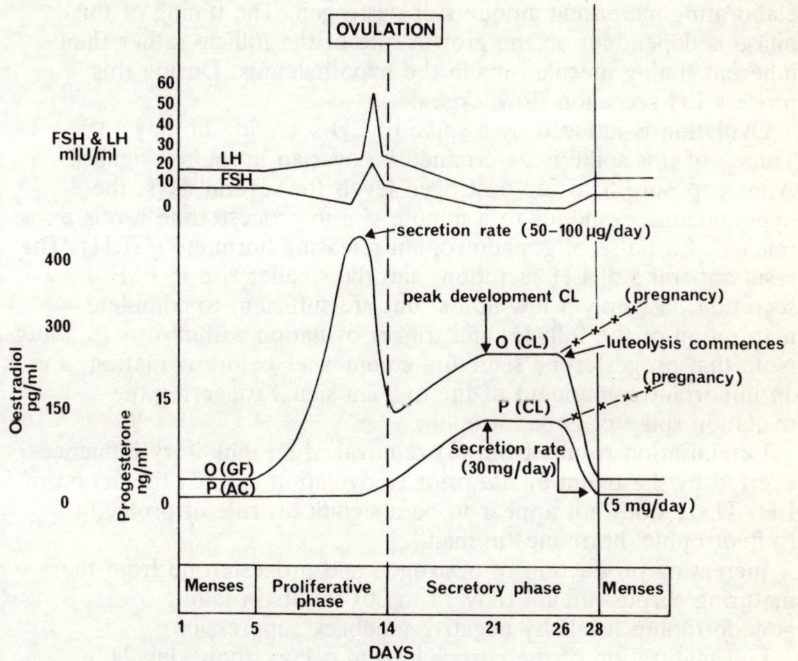

Fig. 6.1 Plasma hormone levels during the menstrual cycle

Key: O = oestradiol GF = graafian follicle CL = corpus luteum
P = progesterone AC = adrenal cortex

The timing of events in the menstrual cycle is determined by the interplay of pituitary and ovarian hormonal signals (see Fig. 6.1).

During the latter portion of the secretory phase FSH and LH secretion are suppressed by the high plasma levels of oestrogen and progesterone. With the withdrawal of oestrogen and progesterone following luteolysis (day 26 to 28) there is a resultant modest rise in FSH (mainly) and LH secretion.

Primary follicles (10 to 15) now commence development under the influence of FSH (day 26 or 27). The development of a single follicle to maturation, with the regression of 9 to 14 others depends upon two facts: (1) after a critical stage of development, about the time the follicular production of oestrogen becomes significant, follicular maturation becomes independent of gonadotrophin; and (2) the first follicle to reach this stage turns off FSH secretion by the negative feedback effect of its secreted oestrogen, thus depriving the remaining follicles, which are still gonadotrophin-dependent, of the stimulus for their development. The single dominant follicle then undergoes full development into a mature Graafian follicle, elaborating increasing amounts of oestrogen. The timing of this phase is dependent on the growth rate of the follicle rather than inherent timing mechanisms in the hypothalamus. During this process LH secretion slowly rises.

Ovulation is induced by a spike of LH secretion on day 13. Timing of this spike is determined by ovarian hormonal signals. After exposure to rising oestrogen levels for several days, the hypothalamus responds to a minute rise in progesterone levels by release of a pulse of gonadotrophin releasing hormone (GRH). The resultant spike of LH secretion, and the smaller rise in FSH secretion, last only a few hours, but are sufficient to complete maturation of the follicle, and trigger ovulation within 6 to 18 hours. Note that progesterone secretion commences before ovulation, and is an important component of the ovarian signal triggering the ovulation spike of gonadotrophin.

Luteinisation results from (1) removal of an inhibitory influence exerted by the ovum *in situ* prior to ovulation and (2) the action of LH. There does not appear to be a significant role of prolactin ('luteotrophic' hormone) in man.

Increasing production of oestrogen and progesterone from the maturing corpus luteum (day 14 to 20) results in falling gonadotrophin levels by negative feedback suppression.

Full maturation of the corpus luteum occurs about day 21. Maintenance of the corpus luteum until day 26 when luteolysis commences, does not seem to be dependent on continued secretion of gonadotrophins. Luteolysis may be prevented by HCG produced by an implanting blastocyst. No role of prostaglandins or trophic hormones has been established in the mechanism of luteolysis in man.

Changes in the ovarian follicle and steroidogenesis during the menstrual cycle

Oestrogen (principally oestradiol) is formed by a cooperative process involving the theca interna cells and the granulosa cells of the developing follicles.

In the early phase of development of the Graafian follicles, the thecal cells (which possess LH receptors) respond to the low LH blood level and synthesise androgen (mainly androstenedione): this is aromatised into oestrogen in the granulosa cells. During this period the granulosa cells of the developing follicles are FSH dependent (possess FSH receptors and few LH receptors) and the aromatising process is potentiated by oestrogen. Oestrogen synergises with FSH; this leads to a rapid proliferation of granulosa cells; an increase in the total population of FSH receptors; and the initial rise in the blood oestrogen level.

This process of steroidogenesis by the thecal cell-granulosa cell unit of the dominant, autonomous follicle continues to increase until day 12 when the oestradiol blood level reaches a preovulatory peak.

By about day 8, the population of FSH receptors on the granulosa cells has been considerably reduced and the cells become insensitive to FSH; from here on the population of LH receptors on the granulosa cells increases and reaches a maximum at the time of ovulation.

Immediately preceding ovulation, LH stimulates the granulosa cells and thecal cells of the dominant follicle to produce progesterone. This leads to the initial small rise in progesterone in the blood which, with the peak level of blood oestrogen, stimulates a pulsatile release of LH, and ovulation follows.

After ovulation, LH stimulates the transformed granulosa and thecal cells to produce oestrogen, progesterone and prostaglandin $F_{2\alpha}$ (an effect which is enhanced by prolactin).

2. Endometrial cycle

a. Proliferative phase

Oestrogen (GF) induces proliferative changes in endometrium (stromal cells, glands and blood vessels).

b. Secretory phase

Oestrogen (CL) increases proliferative changes, and progesterone (CL) causes secretory reactions (glands become tortuous and secretory substances are synthesised: glycogen, lipids and nutrients increase in stromal cells as substrates for implanting blastocyst; and blood flow increases).

The secretory phase only occurs when there has been a normal ovulation. In management of infertility an endometrial biopsy may be taken premenstrually. Histological examination will demonstrate whether secretory phase tissue is present and ovulation has occurred.

c. Menstruation

This follows decreased secretion of oestrogen and progesterone.

Spiral arteries perfuse functional layer which is shed in menstruation. Sequence of events in menstruation:

(i) Constriction of spiral arteries (hours) → ischaemia and necrosis of superficial layer of endometrium.

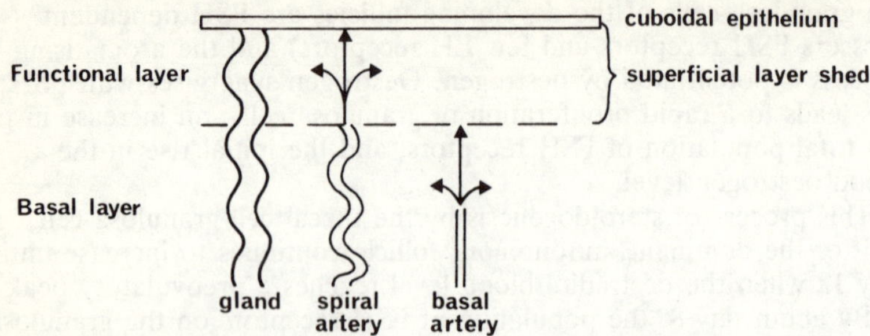

Fig. 6.2 Endometrium

(ii) Dilatation of spiral arteries (minutes) → desquamation, and arterial haemorrhage; vessels then constrict and shut off blood flow.
(iii) Each spiral artery sequentially constricts — dilates — constricts → mininal blood loss (~50 ml) (fibrinolytic and anticoagulant activities in uterus prevent formation of gross blood clots).

3. Cyclical changes in accessory sex organs
Cyclical changes also occur in cervical secretion, vaginal epithelium, Fallopian tubes, breasts and basal body temperature.

Many women can define their date of ovulation by observing the increase in amount of vaginal mucus occurring at this time. Daily measurement of basal body temperature is used clinically to indicate ovulation. Since the rise in temperature only occurs 24 hours *after* ovulation, this test is of little help in prospective indication of the date. Vaginal smears also provide useful information on ovarian hormone secretion.

VAGINAL pH

At the time of ovulation, the acid medium of the vagina (~pH3) is highly spermicidal. During intercourse the pH rises and enhances the motility and viability of ejaculated sperm. Concurrently, the profuse, watery, slightly alkaline cervical mucus assists sperm penetration. This change in vaginal pH is due to:

1. Transudation of fluid from blood in the congested venous plexus surrounding the vagina. The transudate has a pH similar to plasma; its volume increases with the degree of sexual activity; and the vaginal pH may rise to 5 or 6 before ejaculation occurs.
2. Buffer systems (phosphate/bicarbonate) in the ejaculate which rapidly increase the vaginal pH to about 7; even several hours later the pH may be 5 or 6.

OESTROGENS

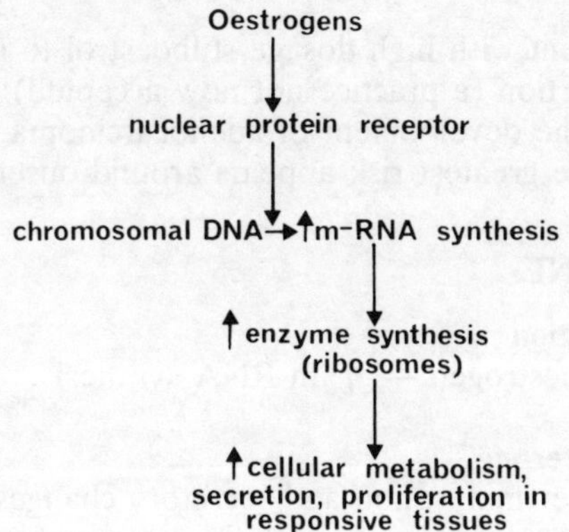

Fig. 6.3 Mechanism of action of oestrogens

Effects of oestrogens

1. Uterus. Epithelium, Endometrial glands, Stromal cells, Myometrium — ↑ proliferation and ↑ vascularity

   ↑ sensitivity to oxytocin, and ↑ spontaneous electrical activity; ↑ cervical mucus — alkaline, copious, watery, minimal viscosity enhances sperm penetration (fern pattern crystallisation on slide).
2. Vagina. Proliferation and cornification of epithelium and glycogen deposition (lactobacilli break down glycogen → acid production → vaginal pH~3 → ↓ viability of pathogenic microorganisms and sperm).
3. Fallopian tubes. Proliferation of mucosa (and muscle); ↑ secretion and ↑ motility of muscle and cilia propel ovum to uterus.
4. Breasts. ↑ duct growth.
5. Other effects. ↓ sebaceous gland secretion (opposes effect of androgen in acne);
   ↑ osteoblastic activity → ↑ rate bone growth, and then → epiphysial closure in long bones → cessation of linear growth;
   ↑ bone matrix (protein); ↑ Ca and Pi deposition;
   NaCl and H_2O retention (via renal tubules);
   ↑ muscle and bone strength at puberty;
   ↑ number of progesterone receptors in uterus;
   In pregnancy or oral contraceptive therapy → suppression of ovulation.

STILBOESTROL AND ADENOCARCINOMA OF THE VAGINA

Maternal treatment with high dosage stilboestrol to prevent spontaneous abortion (a practice not now accepted) has been associated with the development of adenocarcinoma of the vagina in offspring, and the greatest risk appears around onset of puberty.

PROGESTERONE

Mechanism of action
Basically as for oestrogen → ↑ m-RNA synthesis.

Effects of progesterone

1. Uterus. Further maturation and secretory changes in endometrium; needs previous priming action of oestrogen; prepares uterus for implantation of blastocyst;
   ↓ sensitivity to oxytocin, ↑ resting membrane potential and ↓ spontaneous electrical activity (may be used in pregnancy to reduce uterine activity);
   ↓ cervical mucus secretion which becomes viscid and cellular → ↓ sperm penetration.
2. Vagina. Further maturation and proliferation of epithelium; (there are no mucus secreting elements in vaginal epithelium); desquamated cells clump and fold; glycogen content diminishes; leucocytic infiltration occurs; pH rises (4 to 5).
3. Fallopian tubes. Secretory changes in mucosa (for nutrition of fertilised ovum).
4. Breasts. Proliferation of lobule — alveolar system.
5. Other effects. Thermogenic → ↑ basal body temperature (? due to progesterone or aetiocholanolone derivative acting on thermoregulatory centre in hypothalamus); weak antagonism of aldosterone → ↑ NaCl and H_2O loss in urine (competitive inhibition); relaxes smooth muscle in ureters (especially in pregnancy, lesser extent during use of oral contraceptive); relaxes smooth muscle in veins → increased venous pooling of blood; relaxes smooth muscle in arterioles → decreased peripheral resistance (especially in pregnancy).

TRANSPORT AND METABOLISM

Oestrogens
Largely bound to TeBG (testosterone-oestradiol binding globulin) and albumin, also to RBC, very little free in plasma. Oxidised in liver to metabolites or conjugated forming (water soluble) glucuronides and sulphates which are excreted in urine. Some enterohepatic circulation occurs. Diethylstilboestrol and hexoestrol are potent synthetic oestrogens.

Progesterone

Largely bound to albumin and CBG (corticosteroid binding globulin), very little free in plasma. Converted into pregnanediol in liver, conjugated with glucuronic acid and excreted in urine (~10%); the remainder (~90%) is changed into inactive metabolites. Medroxyprogesterone acetate and norethynodrel are potent synthetic progestogens.

FERTILITY DRUGS

An important advance in the treatment of infertility has been the ability to stimulate ovulation in anovulatory women. Clomiphene citrate was first used and probably acts by blocking the effect of oestrogen at receptor sites in the hypothalamus. It is given orally and its most common side effect is the formation of ovarian cysts. The other important ovarian stimulant is human pituitary gonadotrophin, extracted from either postmenopausal urine or human pituitary glands obtained at autopsy. This substance is given by injection and ovarian response is monitored by daily oestrogen assays. Complications include hyperstimulation of the ovaries, and multiple pregnancies. Gonadotrophin releasing hormone is now available, but present indications are that it must be used in conjunction with other ovarian stimulants in order to bring about ovulation.

GRH analogues have been synthesised: some are more potent than GRH (and may become useful in cases of infertility); same competitively inhibit GRH (and may become useful in contraception).

ORIGIN OF TESTOSTERONE IN WOMEN

In normal women, plasma testosterone (range 20 to 70 ng/100 ml or 0.5 to 2.4 nmoles/l) is derived from 2 sources:

1. 50% is secreted into the blood from the adrenal cortex (major source) and from the ovary;
2. 50% is formed from androstenedione (major source), androstenediol, and from dehydroepiandrosterone. These precursors of testosterone are secreted by the adrenal cortex (major source) and the ovary, and are converted into testosterone in the liver (mainly) and peripheral tissues (including sebaceous glands and hair follicles).

In about half the cases of idiopathic hirsutism in women, the plasma testosterone level is elevated. Cyproterone acetate (an androgen antagonist), alleviates hirsutism in a proportion of cases.

Normally in women, secondary sexual hair is dependent on adrenal (not ovarian) function, and androstenedione (not testosterone) is the major precursor of dihydrotestosterone. Androgens maintain sexual drive.

ORIGIN OF OESTRADIOL IN MEN

In men, plasma oestradiol (range 5 to 30 pg/ml or < 100 pmol/l aged 20 to 30) is derived from 2 sources:

1. 30% is secreted by the testis (Leydig/Sertoli cells);
2. 70% is formed from testosterone (major source) and from androstenedione (which is transformed into oestrone). These precursors, which are secreted predominantly by the testis, are converted into oestradiol in extragonadal tissues.

Oestrogen influences testosterone metabolism. It increases the synthesis of testosterone binding globulin (TeBG) (and hence decreases the concentration of free, biologically active, plasma testosterone). Oestrogen also inhibits the conversion of testosterone to dihydrotestosterone.

AMENORRHOEA

Primary: delayed puberty.
Secondary: previously normal menstrual function; excludes anatomical abnormalities (e.g. ovarian agenesis, imperforate vagina, intersexuality).

Amenorrhoea without hypogonadism
Physiological: pregnancy, lactation.
Functional: stress, starvation, psychogenic causes.
Pathological: anatomical abnormalities, Chiari-Frommel syndrome.

Amenorrhoea with hypogonadism
Ovarian failure: climacteric, Turner's syndrome (plasma FSH and usually LH are elevated).

Hypothalamic-pituitary failure: Sheehan's syndrome, prolonged oral contraceptive therapy, phenothiazines (plasma gonadotrophin levels depressed).

Abnormal sex-steroid production: Stein-Leventhal or polycystic ovary syndrome (plasma FSH low; LH increased but lack ovulatory surges), adrenogenital syndrome (plasma gonadotrophin levels depressed), granulosa/theca lutein cysts (plasma gonadotrophin levels depressed).

FEMALE SEXUAL RESPONSE

During intercourse intense venous congestion (due to parasympathetic stimulation) occurs in the labia majora, labia minora, clitoris and vagina. Transudation through the vaginal epithelium from the congested venous plexus provides the lubricant fluid, along with mucus secreted from the cervical glands and a tiny contribution from Bartholin's and Skene's glands.

Increased tone/contraction of the bulbospongiosus and ischiocavernosus muscles (which insert into clitoris) compress the outflow of blood from the clitoris and thereby (with dilatation of clitoral arteries) increase its size (mechanism similar for penile erection). The labia minora are attached to the clitoris and their traction by penile thrusts enhances clitoral excitation. Voluntary contractions of the bulbospongiosus and ischiocavernosus muscles may also stimulate the clitoris.

The vagina becomes more distensible; lubricated by fluid transudation (and to a lesser extent by mucus from bulbourethral glands and urethral glands of penis) and, preceding orgasm, its outer third becomes intensely engorged which decreases the introital lumen.

Progressive elevation of the uterus increases the transcervical diameter of the vagina.

During orgasm, rhythmic contractions of perineal/pelvic muscles occur, especially of the bulbospongiosus ('sphincter vaginae'), ischiocavernosus, sphincter of membranous urethra, transverse/deep perineal muscles, levator ani, and of the uterus and Fallopian tubes. The 5 to 15 contractions of the perineal/pelvic muscles have a rhythm pattern similar to those in muscle groups during ejaculation in the male.

Uterine contractions commencing during orgasm are believed to facilitate sperm transport, although the pattern of uterine contraction (similar to that in labour) suggests an expulsive action. Oxytocin released during intercourse, and seminal prostaglandins, increase uterine contractility.

During the resolution phase of intercourse, the cervix descends into its original position before the expanded inner two-thirds of the vagina assumes its unstimulated state, and hence the cervix is quickly immersed in the constrained seminal pool.

OVARIAN FUNCTION TESTS (PRINCIPLES ONLY)

1. Gonadotrophins

a. Measurement of gonadotrophins
In 24 hour urinary collections or in plasma samples. A single test in a patient who is not ovulating shows whether levels are in the normal range for functioning ovaries or are in the much higher range of the postmenopausal state.

Daily plasma assays of FSH and LH during a menstrual cycle can be performed to demonstrate presence (or absence) of preovulatory increase in LH (up to 60 mIU/ml) and FSH (up to 20 mIU/ml).

b. Clomiphene test
Clomiphene administered orally for five days induces a small rise in plasma gonadotrophins; the test is frequently unreliable.

c. Gonadotrophin releasing hormone test

Synthetic GRH causes LH and FSH release. Measurements are usually made within a 6 to 8 hour period following a subcutaneous injection of releasing hormone.

2. Ovarian hormones

a. Measurement of urinary oestrogens

Daily urinary excretion of total oestrogen is measured, peak values occur 24 hours preovulatory (50 to 100 μg/24 hours) and in luteal phase (40 to 80 μg/24 hours). Useful information can be obtained from even a single sample.

b. Measurement of plasma oestradiol

Peak values occur 24 hours preovulatory (250 to 400 pg/ml; 1000 to 1600 pmol/l) and in luteal phase (100 to 200 pg/ml; 400 to 800 pmol/l).

c. Measurement of urinary pregnanediol

Preovulatory values low (<2.5 mg/24 hours; <8 μmol/24 hours), increasing in luteal phase (2.5 to 10 mg/24 hours; 8 to 30 μmol/24 hours).

d. Measurement of plasma progesterone

Low values (0.5 to 2 ng/ml; <5 nmol/l) before ovulatory LH peak, and peak values (10 to 20 ng/ml; <20 nmol/l) in luteal phase.

3. End organ tests

These include examination of vaginal smears ,endometrial biopsy, cervical mucus preparations, and give valuable information.

FURTHER READING

Barraclough, C. A., Wise, P. M. (1982) The role of catecholamines in the regulation of pituitary luteinising hormone and follicle-stimulating hormone secretion. *Endocrine Reviews*, **3**, 91.

Beyer, C. (ed.) (1979) *Endocrine Control of Sexual Behaviour*. New York: Raven.

Fishman, J. (1976) The cathechol estrogens. *Neuroendocrinology*, **4**, 363.

Fleming, H., Blumenthal, R., Gurpide, E. (1982) Effects of cyclic nucleotides on estradiol binding in human endometrium. *Endocrinology*, **111**, 1671.

George, F. W., Wilson, J. D. (1978) Conversion of androgen to estrogen by the human fetal ovary. *Journal of Clinical Endocrinology and Metabolism*, **47**, 550.

Hersey, R. M., Williams, K. I. H., Weisz, J. (1981) Catechol estrogen formation by brain tissue: characterisation of a direct product isolation assay for estrogen-2- and 4-hydroxylase and its application to studies of 2-and 4-hydroxyestradiol formation by rabbit hypothalamus. *Endocrinology*, **109**, 1912.

Kirschner, M. A., Zucker, I. R., Jespersen, D. (1976) Ovarian and adrenal vein studies in women with idiopathic hirsutism. In James, V. H. T., Serio, J., Giusti, G. (eds) *Endocrine Function of the Human Ovary*, p. 443–455. New York: Academic Press.

Knecht, M., Catt, K. J. (1982) Induction of luteinizing hormone receptors by adenosine 3′, 5′-monophosphate in cultured granulosa cells. *Endocrinology*, **111**, 1201.

Koos, R. D., Clark, M. R. (1982) Production of 6-keto-prostaglandin $F_{1\alpha}$ by rat granulosa cells in vitro. *Endocrinology*, **111**, 1513.

Odell, W. D. (1978) Physiology of the reproductive system in women. In DeGroot, L. J. (ed.) *Textbook of Endocrine Physiology*. New York: Grune & Stratton.

Ojeda, S. R., Campbell, W. B. (1982) An increase in hypothalamic capacity to synthesise prostaglandin E_2 precedes the first preovulatory surge of gonadotropins. *Endocrinology*, **111**, 1031.

Paul, S. M., Axelrod, J. (1977) Catecholestrogens: presence in brain and endocrine tissues. *Science*, **197**, 657.

Rebar, R. W. (1978) Practical evaluation of hormonal status. In Yen, S. S. C., Jaffe, R. B. (eds) *Reproductive Endocrinology*, p. 469. Philadelphia: Saunders.

Ross, G. T., Lipsett, M. B. (eds) (1978) *Reproductive Endocrinology*, *Clinics in Endocrinology and Metabolism*, vol. 7. Philadelphia: Saunders.

Yen, S. S. C. (1978) The human menstrual cycle (integrative functions of the hypothalamic-pituitary-ovarian-endometrial axis). In Yen, S. S. C., Jaffe, R. B. (eds) *Reproductive Endocrinology*. Philadelphia: Saunders.

Multiple choice questions

1. Oestrogen (e.g. oestradiol):
1. increases response of uterus to oxytocin;
2. is synthesised by cells in Graafian follicles but not by cells in the corpus luteum;
3. is secreted into the blood by cells in the testis;
4. can bring about a decrease in FSH secretion;
5. level increases in blood during 24 hours prior to onset of menstrual bleeding;
6. increases the volume and decreases the viscosity of cervical mucus for several days preceding ovulation.

2. In female reproductive disorders:
1. approximately 20% of women who take combined oral contraceptives subsequently develop prolonged amenorrhoea;
2. the presence of a biphasic pattern in the basal temperature chart provides a useful indication of the occurrence of ovulation;
3. urinary pregnanediol excretion should exceed 2 mg per 24 hours during the normal luteal phase;
4. the pituitary gland shows maximal sensitivity to exogenous LRF at mid-cycle;
5. pituitary gonadotrophins rise more markedly after LRF administration during the early follicular phase than during the luteal phase when progesterone levels are elevated;
6. ovulation induction with clomiphene is followed by multiple pregnancy in 50% of cases.

3. Progesterone:
1. decreases response of uterus to oxytocin;
2. is an intermediate in the synthesis of testosterone by the testis;
3. is an intermediate in the synthesis of oestradiol by the ovary;
4. is an intermediate in the synthesis of aldosterone by the adrenal cortex;
5. induces proliferative changes in the endometrium in the week prior to ovulation;
6. during the menstrual cycle reaches its highest level in the blood about 7 to 10 days after ovulation;
7. induces vaginal secretion of mucus.

4. Hypogonadotrophic hypogonadism:
1. may be due to a deficiency of hypothalamic GRH;
2. may be due to a pituitary deficiency of gonadotrophins;
3. may be due to ovarian or testicular disease;
4. may be diagnosed by the gonadotrophin response to GRH infusion;

5. explains the amenorrhoea accompanying chronic malnutrition;
6. explains the impotence of middle-aged diabetic males.

5. With respect to the functions of the ovary:
1. the thecal cells in the Graafian follicle and the corpus luteum secrete some androgen;
2. FSH plasma level begins to rise when the corpus luteum begins to regress (e.g. day 25) and this is concomitant with the onset of growth of a new group of Graafian follicles;
3. FSH causes complete growth/maturation of an ovarian follicle and it can do this without the assistance of LH;
4. following bilaterial oöphorectomy the cyclic secretion of LH is maintained;
5. following hypophysectomy, when ovulation is induced by FSH and LH the corpus luteum can be maintained by LH (without FSH);
6. the response of the ovarian follicles to gonadotrophins can be assessed by plasma oestrogen assays;
7. LH is the ovulatory hormone;
8. in nonpregnant women, prostaglandin $F_{2\alpha}$ causes luteolysis;
9. in the mature Graafian follicle, the presence of the ovum inhibits luteinisation, and when the ovum is removed luteinisation is initiated even in the absence of LH; however, LH is essential for the development/maintenance of the normal corpus luteum.

6. During sexual intercourse:
1. transudation of fluid into the vagina helps to buffer the vaginal acid medium;
2. when semen is ejaculated, the vaginal pH rapidly approximates neutral, and can be maintained close to neutral for several hours;
3. the ejaculated semen in the vasectomised male contains no prostaglandins, since these are made chiefly by the sperm cells;
4. the position of the uterus is not changed, even during sexual excitement before the penis is inserted;
5. orgasm is less intense following hysterectomy;
6. penile thrusts cause stimulation of the clitoris by producing traction on the labia minora which are inserted into the clitoris;
7. oxytocin secreted endogenously, and prostaglandins introduced in the semen, are postulated to enhance uterine contractions;
8. rhythmic contractions of the Fallopian tubes, uterus and bulbocavernosus muscles occur during orgasm, and as long as the sex act is continued a series of orgasms may occur;

9. semen has a lower pH when ejaculations occur at least daily and hence sexual abstinence is recommended in cases where the male is relatively infertile;
10. with sexual excitement, the vagina elongates and expands before the penis is inserted;
11. the ejaculate from the vasectomised male has a lower pH, since the buffers in semen are largely made by the sperm cells.

7. During the week following ovulation:
1. there is an increased plasma progesterone level;
2. there is an increased basal body temperature;
3. there is reduced volume and increased viscosity of cervical mucus;
4. a secretory type of endometrium is formed;
5. there is a decreased plasma oestradiol level which remains at a low level until onset of menstrual bleeding;
6. secretion of LH ceases;
7. secretion of both LH and FSH ceases.

8. Which is the odd one out?
1. testosterone;
2. oestradiol;
3. progesterone;
4. aldosterone;
5. somatomedin;
6. oestriol.

9. With respect to hormones:
1. the mucous secreting cells of the vagina are stimulated by progesterone;
2. LRF implanted in the appropriate area in the hypothalamus increases the release of endogenous LRF, demonstrating the existence of a positive short-loop feedback system;
3. the rise in plasma progesterone which precedes ovulation triggers the preovulatory surge in plasma LH;
4. in women, testosterone in plasma originates from that secreted by the adrenal cortex and ovary, and that derived by metabolism of androstenedione and dehydroepiandrosterone, especially in the liver;
5. when oestradiol elicits a response in a cell it first reacts with a specific membrane receptor, which then activates the adenyl cyclase — c AMP system.

10. Which one of the following is *not* correct?

1. in early follicular development, androgen is formed by thecal cells and is aromatised into oestrogen by granulosa cells;
2. in late follicular development, androgen is formed and aromatised into oestrogen by granulosa cells;
3. in early follicular development, granulosa cells resemble Sertoli cells in that they possess mainly FSH receptors and function in gametogenesis;
4. following ovulation, FSH induces luteinisation of granulosa and thecal cells, and secretion of oestradiol, progesterone and prostaglandin $F_{2\alpha}$;
5. catecholoestrogens probably stimulate GRH secretion;
6. relaxin is a polypeptide secreted by the ovary which probably reduces uterine contractions and induces softening of uterine cervix.

11. Which of the following are correct?

1. folliculostatin (ovarian inhibin) is produced by granulosa cells and selectively inhibits FSH secretion;
2. an oocyte maturation inhibitor is present in follicular fluid and is made by granulosa cells;
3. both progesterone synthesis and prostaglandin $F_{2\alpha}$ synthesis by follicular cells are necessary for rupture of the follicle;
4. intrauterine devices prevent fertilisation by setting up sterile inflammatory processes in the endometrium and inducing release of lysosomal enzymes from leukocytes;
5. clomiphene acts by inhibiting inhibitory influences on GRH release;
6. oestrogens increase the hepatic synthesis of some hormone carrier proteins, clotting factors (II, VII, IX, X), angiotensinogen, very low density lipoproteins and high density lipoproteins.

Answers

1. 1, 3, 4, 6
2. 2, 3, 4
3. 1, 2, 3, 4, 6
4. 1, 2, 4, 5
5. 1, 2, 5, 6, 7, 9
6. 1, 2, 6, 7, 8, 10
7. 1, 2, 3, 4
8. 5
9. 3, 4
10. 4
11. 1, 2, 3, 4, 5, 6

---

7. Intermediary metabolism

CARBOHYDRATE METABOLISM (PRINCIPLES)

With respect to carbohydrate metabolism the following brief outlines are important (see Fig. 7.1):

1. In muscle, fat cell and nongluconeogenic tissue
Irreversible reactions in glycolysis:

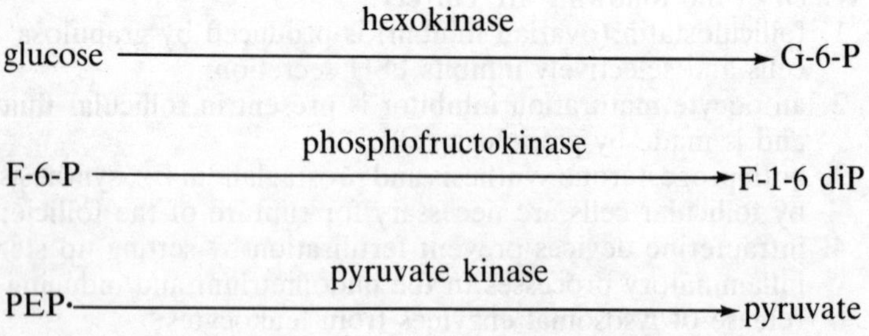

2. In gluconeogenic tissue — liver (major organ) and kidney
The above reactions can be reversed so that both glucose synthesis and degradation are possible:

G-6-P —(phosphatase)→ glucose

F-1-6 diP —(FDP-ase)→ F-6-P

pyruvate + ATP + CO_2 —(pyruvate carboxylase)→ oxaloacetate + ADP + P_i

oxaloacetate + GTP —(PEP carboxykinase)→ PEP + CO_2 + GDP

These enzymes are low/absent in other tissues.

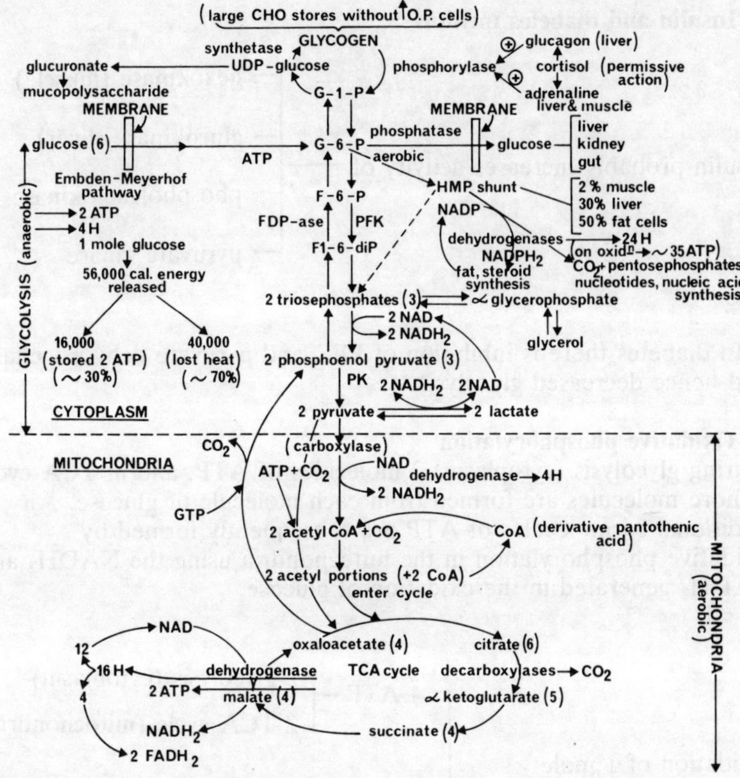

Fig. 7.1 Principles of carbohydrate metabolism

Key: triosephosphates = phosphoglyceraldehyde (glyceraldehyde-3-phosphate) and dihydroxyacetone-P.

HMP = hexose monophosphate shunt (aerobic)
GTP = guanosine triphosphate
UDP- = uridine diphospho- (as in UDP-glucose)
CHO = carbohydrate
PK = pyruvate kinase
FDP-ase = fructose diphosphatase
PEP = phosphoenolpyruvate
PEP carboxykinase = phosphoenolpyruvate carboxykinase
PFK = phosphofructokinase
NAD = nicotinamide adenine dinucleotide (coenzyme I) (DPN)
NADP = nicotinamide adenine dinucleotide phosphate (coenzyme II) (TPN)
$NADH_2$ = dihydro (form above)
$NADPH_2$ = dihydro (form above)
(2), (3) etc. = number of carbons in compound

3. Insulin and diabetes mellitus

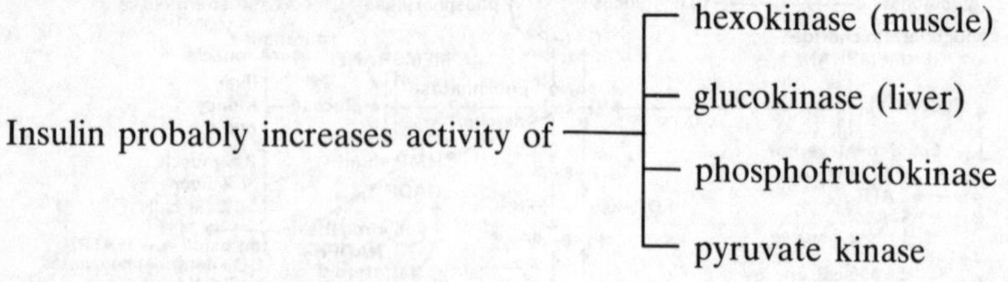

In diabetes there is inhibition of PFK and pyruvate dehydrogenase and hence decreased glycolysis.

4. Oxidative phosphorylation
During glycolysis (cytoplasm) 2 molecules of ATP, and in TCA cycle 2 more molecules are formed from each molecule of glucose. An additional 34 molecules of ATP are subsequently formed by oxidative phosphorylation in the mitochondria using the $NADH_2$ and $FADH_2$ generated in the oxidation of glucose.

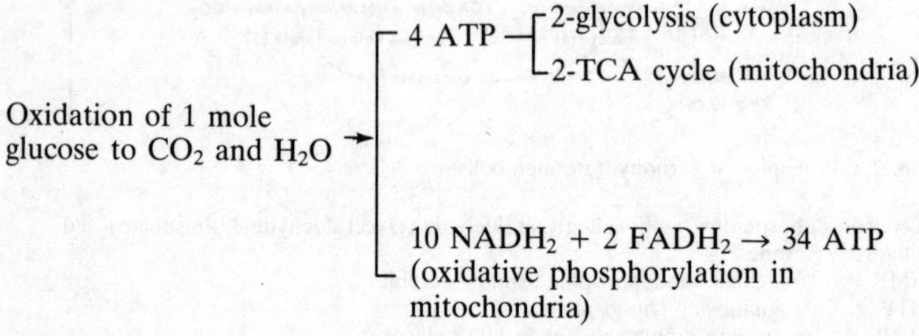

(When pyruvate is formed from glycogen 1 mole G-6-P → 3 ATP.)

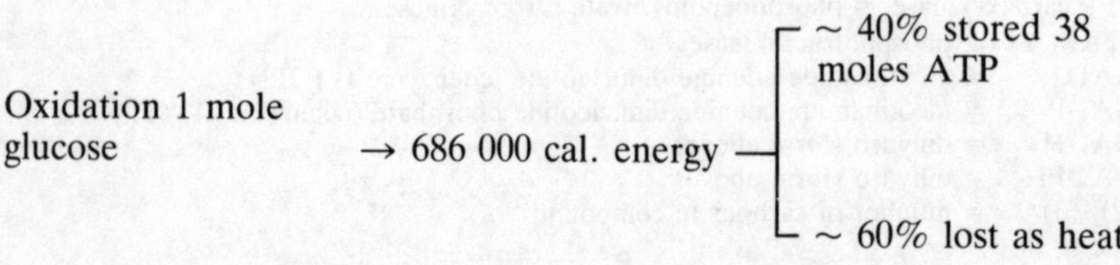

The enzymes for glycolysis in the Embden-Meyerhof pathway are present in the cytoplasm.

5. Role of ATP

$$1\ \text{mole ATP} \underset{+8000\ \text{cal}}{\overset{-8000\ \text{cal}}{\rightleftarrows}} 1\ \text{mole ADP}\ (\text{physiological conditions})$$

ATP provides energy for (biological work)
- muscle contraction
- cell secretion, synthesis
- active transport → membrane potentials

About 60% of energy in glucose and fat (FFA) is lost as heat during oxidation, and ~ 40% is stored in ATP. When ATP releases energy to functional systems of cells still more energy is lost as heat, so that only about 25% of all energy in food is used for functional purposes and 75% is dissipated as heat.

6. Liver-blood glucose regulation

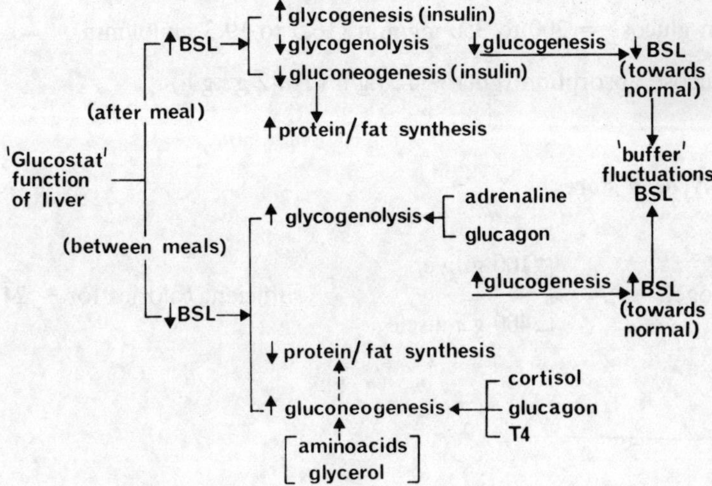

Fig. 7.2 Glucostat function of liver

The BSL is maintained within the range 70 to 120 mg/100 ml (3.8 to 6.6 mmol/l): glucose is normally the only metabolic substrate for nerve cells.

After a meal the BSL rises, and it would increase considerably more without buffering effect of liver (and insulin). During interdigestive periods, BSL is maintained relatively steady, especially by gluconeogenic function of liver and by glycogenolysis (otherwise ↓ BSL → deleterious effects on neuronal and retinal function, and white blood cell production). Skeletal muscle stores and utilises glycogen, however FFA/ketone bodies are preferentially oxidised thus ensuring adequate glucose supply to brain.

7. Fate of glucose absorbed from small intestine

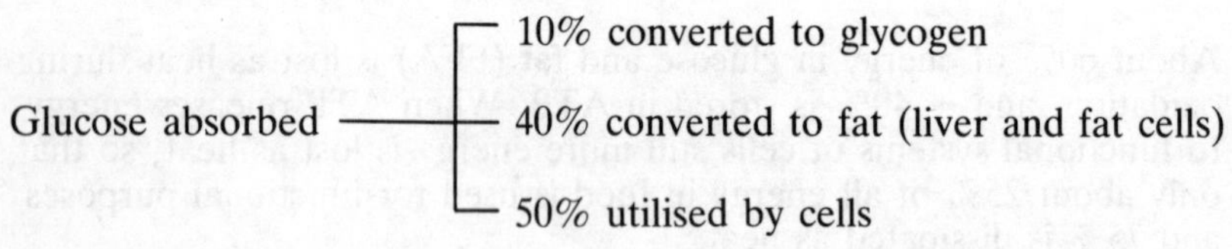

Glucose absorbed in excess of that utilised by cells or converted to glycogen is stored as fat in fat cells and lesser extent in liver.

Tm glucose = 300 to 350 mg/min (16.5 to 19.3 mmol/min)

Glucose absorption (gut) ~ 120 g/h (Tm 2 g/kg/h)

8. Glycogen stores

Glycogen ─┬ 100 g liver
          └ 400 g muscle
→ sufficient calories for ~ 24 h

BASAL ENERGY REQUIREMENTS AND FUEL RESERVES

Basal daily energy turnover and fuel reserves for a 70-kg adult male:

Basal daily energy turnover
- carbohydrate: 200 g; 3435 kJ; 820 kcal; *49%
- protein: 70 g; 1215 kJ; 290 kcal; *17%
- fat: 60 g; 2413 kJ; 576 kcal; *34%

Fuel reserves
- fat 80% (fat → 40 kJ/g; 9.6 kcal/g)
- carbohydrate < 0.5% (carbohydrate → 17 kJ/g; 4.1 kcal/g)
- protein 20% (protein → 17 kJ/g; 4.1 kcal/g)

BIOLOGICAL OXIDATION

See Figure 7.3.

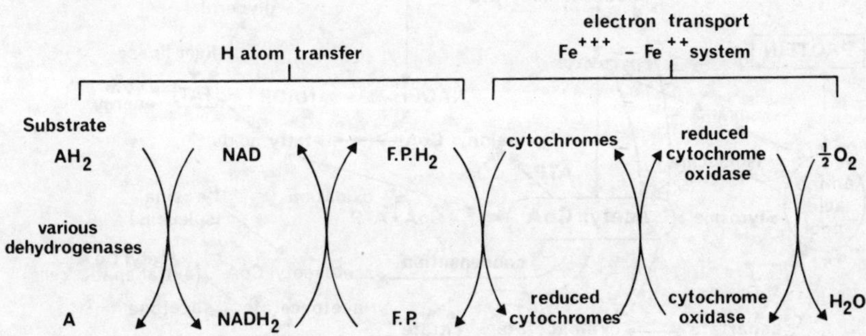

Fig. 7.3 Mitochondrial electron transport chain

* = % of total

Oxidative phosphorylation = formation of ATP during oxidation of $NADH_2$ or $FADH_2$ (mitochondria).

H atom/electron is removed from substrate (e.g. glucose/FFA) and shuttled sequentially by oxidative enzymes in mitochondria for oxidation.

A hydrogen atom removed during dehydrogenation is ionised as follows:

$$\underset{\text{atom}}{H} \leftrightarrows \underset{\text{proton}}{H^+} + \underset{\text{electron}}{\varepsilon} \rightarrow (Fe^{+++} \rightarrow Fe^{++})$$

Oxidation — H removed or electron lost or O gained

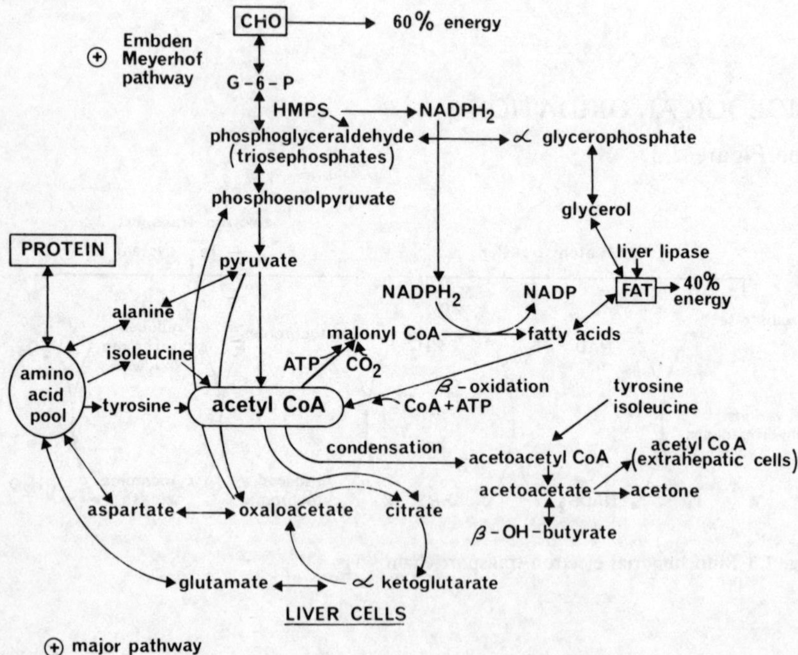

Fig. 7.4 Interrelationship of carbohydrate, protein and fat metabolism

INTERRELATIONSHIP — CARBOHYDRATE, PROTEIN, FAT (PRINCIPLE)

This is illustrated in Figure 7.4.

Interconversions of carbohydrate, protein and fat are simplified as follows:

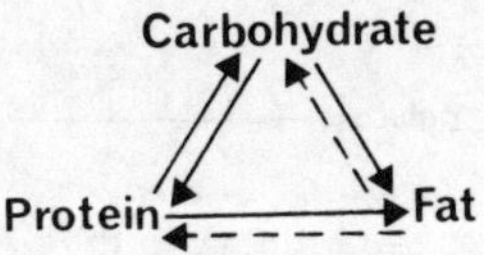

In animals, only the glycerol carbons of the triglycerides can by used to make glucose or be incorporated into the non-essential amino acids.

Lipogenesis = Synthesis of triglycerides from fatty acids and α-glycerophosphate, occurring mainly in liver and adipose tissue (in microsomes).

Lipolysis = Hydrolysis of stored fat to give glycerol and fatty acids. β-oxidation of fatty acids (removal 2C units) (in mitochondria, liver, muscle, etc.).

Normally more acetyl CoA is formed from this source that can be used by liver (rest used by other cells).

Acetyl CoA $\xrightarrow{\text{acetyl carboxylase}}$ malonyl CoA → FFA. In diabetes, acetyl carboxylase is deficient, therefore FFA synthesis is reduced and excess acetyl CoA is converted to ketone bodies.

GLUCOSE-FATTY ACID (FFA) CYCLE IN ADIPOSE TISSUE

This is shown in Figures 7.5a and 7.5b.

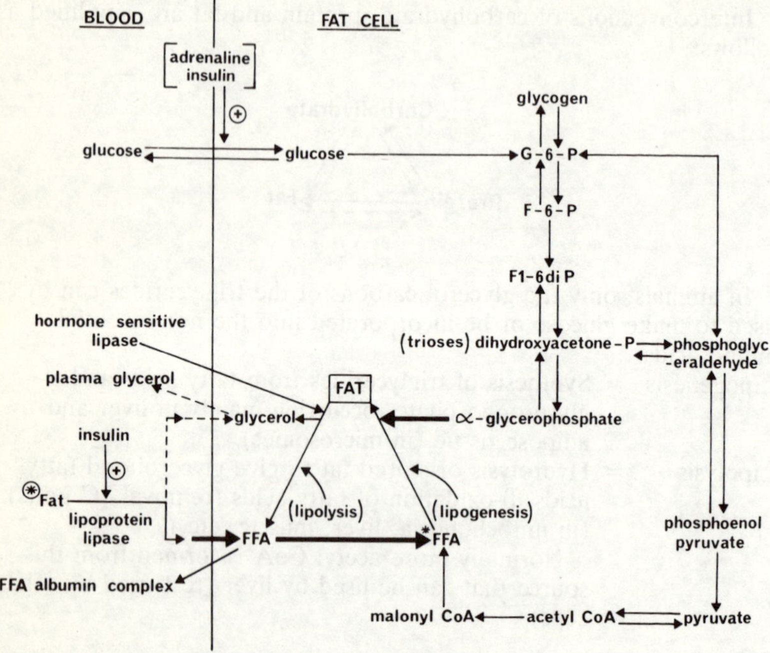

Fig. 7.5a Glucose-fatty acid (FFA) cycle in adipose tissue

Key: ⊛ = Fat in chylomicron and lipoprotein triglyceride form.

* = FFA is activated by ATP/CoA to form fatty acyl-CoA, which reacts with α-glycerophosphate → triglyceride. Glucose → acetyl CoA for *FFA synthesis, and α-glycerophosphate for triglyceride production.

⊞ = Glycerol entering cell is of *minor* importance in triglyceride production. In triglyceride synthesis, FFA entering cell are quantitatively more important than those produced within cell

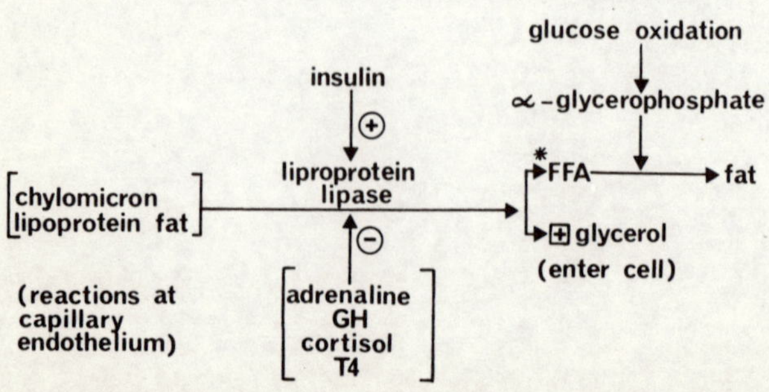

Fig. 7.5b Lipoprotein lipase activity

Note: Glycerol kinase is present in liver to convert glycerol to α-glycerophosphate, and thence to other intermediates/glucose. This kinase is not present in fat cells, the glycerophosphate used coming from glucose — this link between glycolysis and triglyceride synthesis in fat cells is presumably important in control.

FREE FATTY ACID (FFA) MOBILISATION (FROM FAT CELLS); TRANSPORT AND UTILISATION

See Figure 7.6.

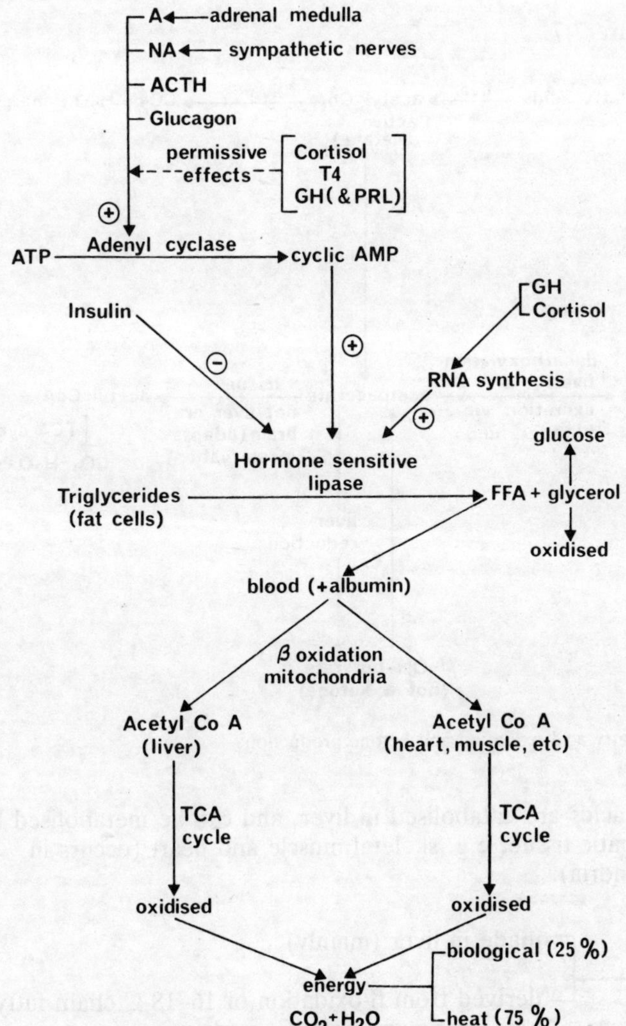

Fig. 7.6 Free fatty acid (FFA) mobilisation (from fat cells); transport and utilisation

There is a reciprocal inhibitory relationship between glucose and fatty acid utilisation. When oxidation of glucose is increased the oxidation of fatty acids is decreased (and vice-versa).

$$\begin{array}{ccc} \text{Glucose oxidation} & \xleftrightarrow{\ominus} & \text{FA oxidation} \\ \downarrow & & \downarrow \\ \sim 50\% & \text{— energy —} & \sim 50\% \end{array}$$

FATTY ACID OXIDATION AND KETONE PRODUCTION

See Figure 7.7.

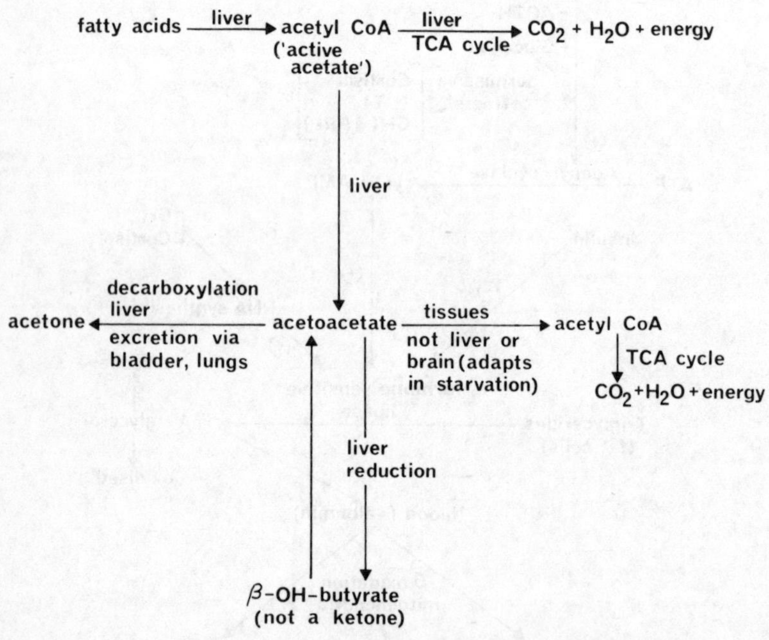

Fig. 7.7 Fatty acid oxidation and ketone production

Fatty acids are catabolised in liver, and can be metabolised in extrahepatic tissue, e.g. skeletal muscle and heart (occurs in mitochondria).

Ketone bodies
- made in liver (mainly)
- derived from β-oxidation of 16–18 C chain fatty acids, e.g. stearic and palmitic acids

The intermediates in the citric acid cycle, including oxaloacetate, can be replenished using amino acids (via glutamate or aspartate), or from pyruvate → oxaloacetate. Therefore carbohydrate metabolism may be deficient (as in diabetes, starvation) but citric acid cycle intermediates can be maintained at normal levels, to allow oxidation of acetyl CoA at a normal rate. In these conditions, acetyl CoA is produced at a greater rate than normal, probably to provide enough glucose from glycerol. The formation of acetoacetate is the only way the liver can dispose of the excess production of acetyl CoA.

GLUCONEOGENESIS

See Figure 7.8.

Gluconeogenesis is formation of glucose from nonglucose precursors, especially amino acids (exception being leucine) and glycerol. Glucose can also be formed from pyruvate, lactate, fructose, and dihydroxyacetone phosphate — in fact, from any of the intermediates of glycolysis and the citric acid cycle. Gluconeogenesis

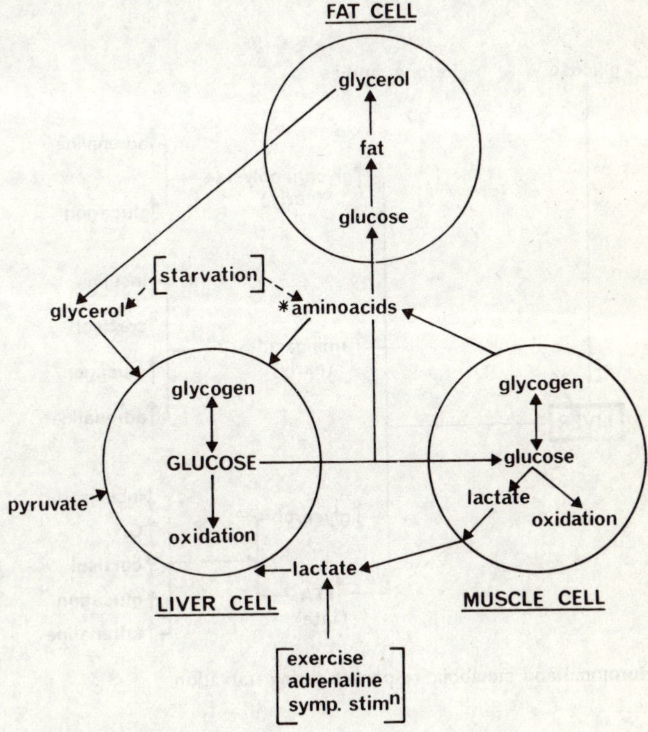

Fig. 7.8 Gluconeogenesis

* = main source

in kidney is quantitatively less important. During fasting (for several days), protein is the most important glucose source, and during exercise, it is lactate.

Gluconeogenesis is important during interdigestive periods (especially at night) to maintain a reasonably steady blood glucose level for brain cell metabolism.

STARVATION

See Figure 7.9.

During starvation the blood glucose level has to be maintained above 60 mg/100 ml (3.3 mmol/l) to ensure an adequate energy source for brain cell activity. In the early stages, this is achieved by increased liver gluconeogenesis, using gluconeogenic amino acids

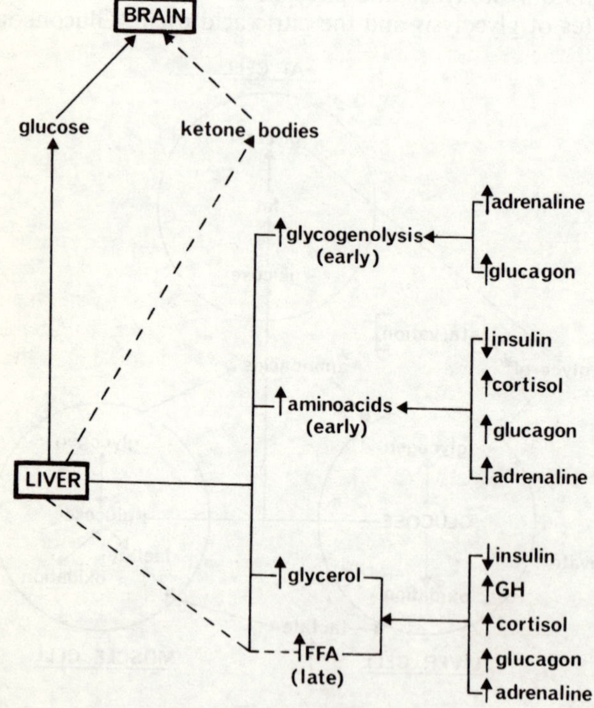

Fig. 7.9 Hormonal and metabolic responses during starvation

(mainly) from muscle, and glycerol (of less importance) from fat cells. As starvation proceeds (beyond a week or so) glucose produced from these sources is supplemented by an increased mobilisation of FFA from fat cells and their conversion into ketone bodies in the liver (FFA cannot be synthesised into glucose). The brain adapts by increasing its utilisation of ketone bodies (acetoacetate/β-OH-butyrate). Although the brain can adapt to using ketone bodies in greater amounts than normal, these supply only about 50% of its energy requirements. Hence, blood glucose concentration must still be maintained.

The secretion of insulin is decreased during fasting, and this accounts significantly for the metabolic responses which follow: other hormones also exert effects.

After fat stores have been exhausted the basal energy needs are supplied by body protein (e.g. in marasmus), and hypoglycaemia becomes severe. Blood glucose concentration initially falls soon after glycogen stores have been utilised (~ 24 hours), but gluconeogenesis maintains glucose level above that which causes hypoglycaemic reactions (these occur when blood glucose concentration becomes <60 mg/100 ml or <3.3 mmol/l).

Note

- brain can normally utilise substrates other than glucose (e.g. fructose, lactate, pyruvate, α-ketoglutarate, oxaloacetate), but insufficient amounts of these are transported across blood-brain barrier for its energy needs
- insulin is not secreted when blood glucose concentration falls below 60 mg/100 ml (3.3 mmol/l); hence, glucose is not utilised by muscle and adipocyte cells, and fat is their main energy substrate
- minimal oxidation of glucose is essential for complete oxidation of FFA in tissues
- newborn babies can tolerate blood glucose concentrations as low as 20 mg/100 ml (1.1 mmol/l); probably because adrenaline increases lactate (from glycogenolysis in muscles), and FFA and glycerol which permeate blood-brain barrier and are utilised by brain

METABOLIC ACTIONS OF ADRENALINE

See Figure 7.10.

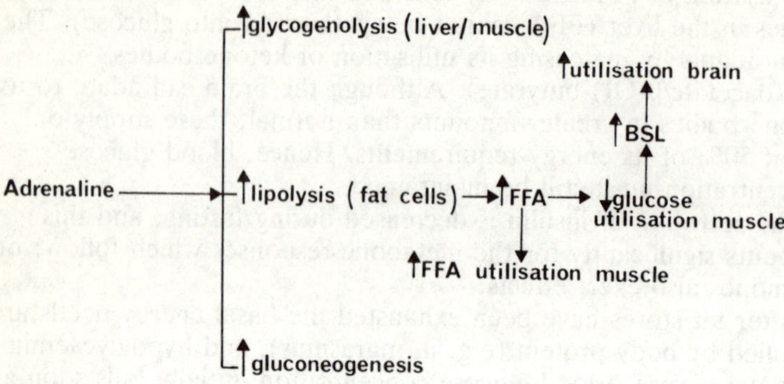

Fig. 7.10 Metabolic actions of adrenaline

FURTHER READING

Albrink, M. J. (1974) Overnutrition and the fat cell. In Bondy, P. K., Rosenberg, L. E. (eds) *Duncan's Diseases of Metabolism*, 7th edn, p. 417. Philadelphia: Saunders.

Alfin-Slater, R. B., Kritchevsky, D. (eds) (1979) *Human Nutrition: A Comprehensive Treatise*. New York: Plenum.

Alfin-Slater, R. B., Kritchevsky, D. (eds) (1979) *Nutrition and the Adult: Macronutrients*. New York: Plenum.

Baer, H. P., Drummond, G. I. (eds) (1979) *Physiological and Regulatory Functions of Adenosine and Adenine Nucleotides*. New York: Raven.

Baltcheffsky, H., Baltscheffsky, M. (1974) Electron transport phosphorylation. *Annual Review of Biochemistry*, **43**, 871.

Butler, T. M., Davies, R. E. (1980) High-energy phosphates in smooth muscle. In Bohr, D. F. et al (eds) *Handbook of Physiology*, section 2, vol. II, p. 237. Baltimore: Williams & Wilkins.

Felig, P., Saudek, C. D. (1976) The metabolic events in starvation. *American Journal of Medicine*, **60**, 117.

Shils, M. E., Goodhart, R. S. (eds) (1979) *Modern Nutrition in Health and Disease*. Philadelphia: Lea & Febiger.

Young, V. R., Scrimshaw, N. S. (1971) The physiology of starvation. *Scientific American*, **225**, 14.

Multiple choice questions

1. During biological oxidation:

1. the substrate being oxidised loses H or an electron or gains O;
2. dehydrogenases remove 2H atoms from their substrates and use these H atoms to reduce the next electron carrier;
3. NAD + 2H → $NADH_2$ (where NAD = coenzyme 1 = DPN);
4. in fat cells, glucose oxidation generates α-glycerophosphate, which reacts with fatty acyl CoA to form triglyceride;
5. in fat cells, FFA which enter following hydrolysis of fat from lipoproteins are activated by ATP/CoA, and react with α-glycerophosphate to produce triglyceride;
6. in fat cells, FFA produced from acetyl CoA which is produced from glucose are quantitatively more important than those which enter the cells;
7. in fat cells, insulin facilitates glucose entry and subsequent lipogenesis;
8. in fat cells, glycerol is of major importance in triglyceride synthesis because glycerol kinase is present.

2. During metabolism:

1. FFA released from fat cells are utilised in liver cells, but not in muscle cells;
2. heart, like brain, utilises glucose almost exclusively;
3. in starvation, brain cells can use ketone bodies for some of their energy needs;
4. lactate, formed from pyruvate in the cytoplasm of muscle cells, can be converted to glucose in the liver;
5. aspartate can be converted into oxaloacetate, and then oxidised via the TCA cycle;
6. the major source of ketone bodies is skeletal muscle;
7. glucose oxidation generates oxaloacetate, which is required (in small amounts) for FFA oxidation in mitochondria;
8. normally, when the oxidation of glucose is increased then the oxidation of FFA is decreased;
9. glucose can be formed from nonglucose precursors, e.g. lactate, amino acids, glycerol and pyruvate; and this gluconeogenesis process takes place mainly in the liver;
10. skeletal muscle and heart utilise FFA and ketone bodies in preference to glucose when glucose is deficient, e.g. in starvation.

3. During carbohydrate metabolism:

1. glucose is transported across muscle cell membranes by carrier-mediated diffusion, a process which is facilitated by insulin;

2. glycolysis via the Embden-Meyerhof pathway is anaerobic and 2 molecules of ATP (net) are produced by the conversion of 1 molecule of glucose to 2 molecules of lactate;
3. glycolysis via the Embden-Meyerhof pathway occurs in the cytoplasm;
4. glucose oxidation via the HMP shunt generates $NADPH_2$ which is involved in steroid/fatty acid synthesis;
5. glucose metabolism via the Embden-Meyerhof pathway is normally of minor importance in muscle cells;
6. when glucose is converted to lactate via the Embden-Meyerhof pathway about 70 per cent of the energy released is lost as heat;
7. aerobic oxidation of pyruvate takes place in mitochondria;
8. oxidation of 1 molecule of glucose by O_2 in cells produces 6 molecules each of CO_2 and H_2O and generates 38 ATP molecules (net).

4. During glucose metabolism:
1. the steps from G-6-P to pyruvate by the Embden-Meyerhof pathway occur in the cytoplasm;
2. conversion of glucose to G-6-P is catalysed by hexokinase in muscle cells;
3. pyruvate is oxidised in mitochondria;
4. lactate produced from glucose in muscle during contraction is converted back to glucose in resting muscle;
5. the mitochondrial membrane is freely permeable to $NADH_2$ produced during glycolysis in the cytoplasm.

5. During metabolism:
1. 1 g of fat when oxidised produces 9 Calories of energy;
2. while fasting for 48 hours, the most important glucose sources for tissues in general are glycogen and amino acids;
3. ketone bodies are produced by the liver and utilised by peripheral tissues as a normal process: over-production of ketone bodies occurs during prolonged starvation, vigorous exercise or fever;
4. the ribosome content of muscle cells is decreased after hypophysectomy, and this results in decreased protein synthesis due chiefly to loss of GH;
5. actinomycin D inhibits m-RNA synthesis;
6. in the liver, amino acids can be transformed into glucose, urea, FFA and protein.

6. During metabolism:
1. insulin facilitates glucose transport across muscle cell membranes;
2. sympathetic stimulation increases and insulin decreases the activity of hormone sensitive lipase;

3. G-6-P is oxidised in part via the HMP shunt (especially in fat cells);
4. the uptake by fat cells of FFA from lipoprotein triglycerides is enhanced by insulin;
5. glucose after entering a skeletal muscle cell can leave as glucose;
6. protein synthesis is increased by insulin and decreased by cortisol;
7. FFA mobilised from fat cells are transported in blood complexed with albumin.

7. During biological oxidation and intermediary metabolism:
1. the process of dehydrogenation may be formally represented as the removal of 2H atoms from substrates;
2. H atoms are transported via NAD (or NADP) coenzymes;
3. H_2O is formed as end product;
4. electron transport occurs via Fe^{+++} — Fe^{++} system in cytochrome enzymes;
5. alanine can be reversibly converted to pyruvate in liver;
6. aspartate can be reversibly converted to oxaloacetate in liver;
7. acetoacetyl-CoA is formed in liver;
8. FFA are oxidised in mitochondria in liver and heart.

8. During starvation:
1. liver glycogen deposits are completely utilised within 30 hours; muscle glycogen and brain glycogen probably are replenished by gluconeogenesis;
2. FFA oxidation in liver, muscle and heart is increased;
3. ketone bodies produced in liver from FFA can be utilised by brain cells but glucose is still essential;
4. there is a fall in total K content of the body.

9. During carbohydrate metabolism:
1. liver but not muscle cells contain glucose-6-phosphatase to allow formation of glucose from glucose-6-phosphate;
2. in muscle and fat cells, pyruvate is not converted to oxaloacetate and thence to glucose;
3. the oxidation of 1 g glucose to $CO_2 \rightarrow$ 4 Calories of energy;
4. in liver cells, the reaction glucose $\rightarrow$ G-6-P can be reversed by phosphatase;
5. the oxidation of 1 mole glucose by O_2 in cells produces 6 moles each of CO_2 and H_2O and generates about 700 000 calories of energy of which about 60% is lost as heat;
6. in muscle cells, in the absence of insulin no glucose enters.

10. With respect to gluconeogenesis:
1. it is not initiated until fasting extends beyond 24 hours;
2. during exercise, lactate is the major gluconeogenic compound;

3. it is decreased by sympathetic stimulation;
4. during starvation, in the first 36 hours amino acids derived from muscle proteins are the major source of new glucose formation;
5. during starvation, after several days FFA mobilised from fat cells are the major source of new glucose formation.

11. Given that an average 70 kg man has about 250 g glycogen in his liver and about 300 g glycogen in his muscle stores, about how many hours would it take to deplete these stores completely if he remained in the basal state (70 C/h)?
1. 10;
2. 20;
3. 30;
4. 40;
5. 50.

12. With respect to utilisation of FFA and ketones: which is the odd one out?
1. liver;
2. muscle (skeletal);
3. kidney;
4. heart;
5. brain;
6. stomach.

Answers

1. 1, 2, 3, 4, 5, 7
2. 3, 4, 5, 7, 8, 9, 10
3. 1, 2, 3, 4, 6, 7, 8
4. 1, 2, 3
5. 1, 2, 3, 4, 5, 6
6. 1, 2, 3, 4, 6, 7
7. 1, 2, 3, 4, 5, 6, 7, 8
8. 1, 2, 3, 4
9. 1, 2, 3, 4, 5
10. 2, 4
11. 3
12. 5

---

8. Pancreas

HISTOLOGY

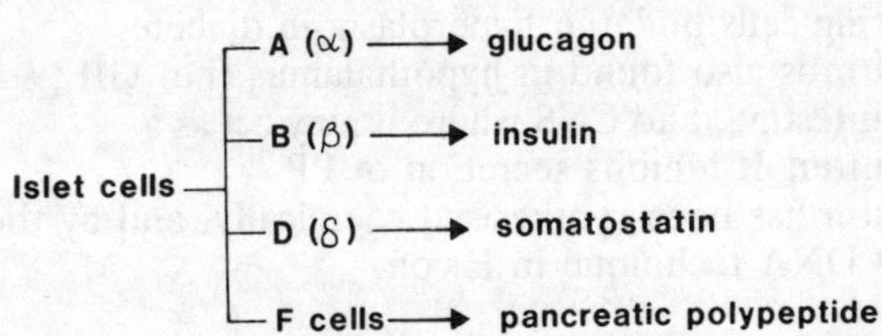

ISLET-CELLS INTERRELATIONSHIPS

The paracrine effects of islet-cells are shown in Figure 8.1.

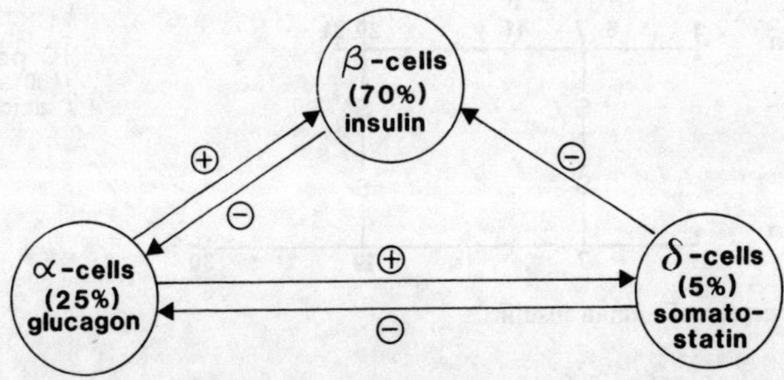

Fig. 8.1 Possible paracrine regulatory action on secretion of islet-cell hormones

% = percentage of cell types in normal islets; in juvenile-onset diabetes, proportion of β-cells is reduced and proportion of α-and δ-cells is increased; in adult-onset diabetes, proportion of δ-cells is low

The islet-cells form functional (chemical) and anatomical cell-cell relationships. Functionally, the islet behaves as a small organ ('islet-as-mini-organ' concept), and excitatory or inhibitory impulses elicit a coordinated response by the integrated activity of its cells. This coordinated response is mediated by the effects of islet-cell hormones on the secretory activity of neighbouring cells (paracrine effects). Anatomically, the cells form either tight junctions or gap junctions (where there is continuity of cytoplasm) between neighbouring cells.

Pancreatic polypeptide (PP) (MW 3600) is produced by islet-cells (F cells) and by endocrine cells outside the islets between the acinar cells. PP inhibits secretion of pancreatic enzymes and contractility of gall bladder (cf. PZ-CCK). PP secretion increases after a meal (cf. insulin), and during hypoglycaemia, fasting and exercise. Normal values <1000 ng/l (240 pmol/l).

PP-producing cells undergo hyperplasia in diabetes.

Somatostatin is also found in hypothalamus (i.e. GIF; see p. 4), small intestine, and CNS where it may act as a neurotransmitter. It inhibits secretion of PP.

Somatostatin has been synthesised chemically, and by the recombinant DNA technique in E.coli.

STRUCTURE OF HUMAN INSULIN

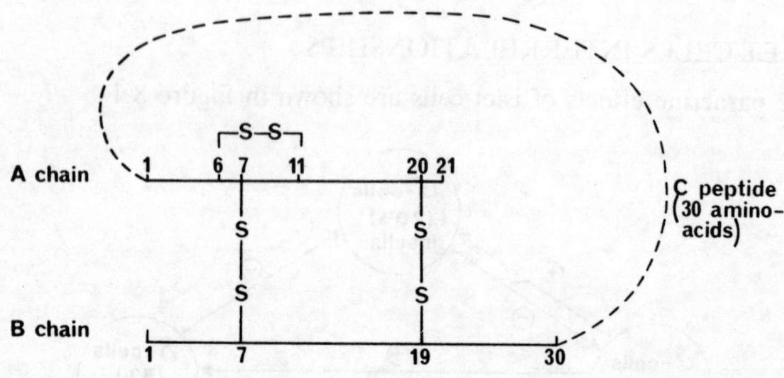

Fig. 8.2 Structure of human insulin

Insulin is a protein, 51 amino acids, MW ~6000 (human); 2 chains linked by disulphide bridges (rupture → inactivity). Species differences occur at amino acids 8, 9, 10 (A chain), 30 (B chain). Synthesised in ribosomes of endoplasmic reticulum of β-cells as proinsulin (81 amino acids), the C peptide is removed by enzyme hydrolysis and secreted in equimolar amounts with insulin (plasma C peptide level 0.5 to 2.5 μg/l). Insulin exists as a monomer, dimer or hexamer (3 dimers). It is stored in granules (hexamers complexed with Zn^{++}) prior to secretion which is initiated by Ca^{++} influx into the cells and onto the microtubular storage system. Insulin has been synthesised by the recombinant DNA technique in E. coli, and in small (non commercial) amounts from its constituent amino acids.

Proinsulin is present in plasma and can be estimated as a component of the total immunoreactive insulin (IRI). IRI is <20 μ units/ml or <20 mIU/l (after overnight fast, normally).

Glucose stimulates entry of Ca^{++} into β-cells, synthesis of proinsulin, and secretion of insulin, C peptide and small amounts of proinsulin. A specific radioimmunoassay for C peptide can be employed to estimate endogenous insulin production in diabetic patients receiving insulin.

When plasma insulin is estimated by bioassay, the result is referred to as insulin-like activity (ILA), and this always exceeds IRI value. ILA includes substances with insulin-like actions which do not react with insulin antibody. When insulin antibody is added the effect of IRI is nullified, and the ILA that persists is known as non-suppressible insulin-like activity (NSILA) (and belongs to the family of somatomedins, see p. 10).

Transport and metabolism
Secretion rate 1 to 2 mg/d.

Probably bound (in part) to plasma proteins; $t_{\frac{1}{2}} \sim 4$ min; inactivated by insulin-glutathione transhydrogenase (liver), and by proteolytic enzymes ('insulinase') present in many tissues, especially liver and kidney. Specificity and significance of these enzyme systems not known; 40% of insulin removed from circulation during single passage through liver.

ACTIONS OF INSULIN

These are initiated by insulin-receptor interactions at cell membranes. Insulin receptors appear to be a complex of 2 or more proteins. In rare instances some insulin-resistant diabetics have antibodies to their own insulin receptors.

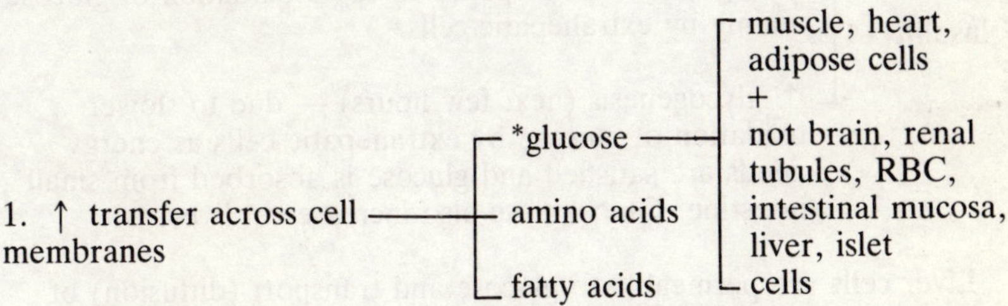

+ = except satiety centre in ventromedial hypothalamus
(Insulin does not cross blood/brain barrier; glucose diffuses across this barrier).

2. ↑ glucose oxidation (insulin sensitive cells — muscle, adipose, heart) (see Figs 7.1, 7.5a).

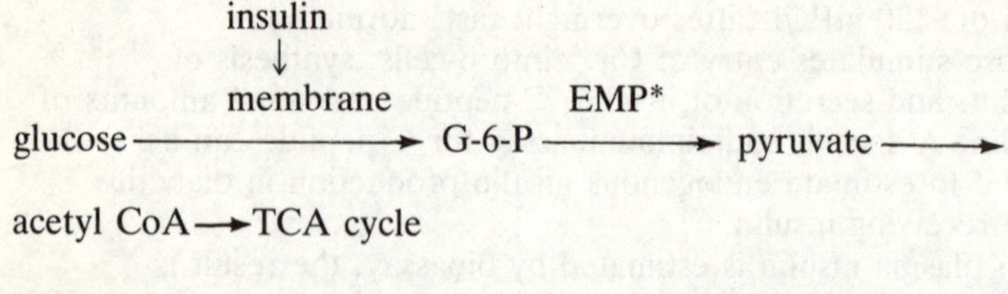

Insulin possibly increases activity of
- hexokinase
- phosphofructokinase
- pyruvate kinase

3. ↑ glycogenesis (mainly muscle)

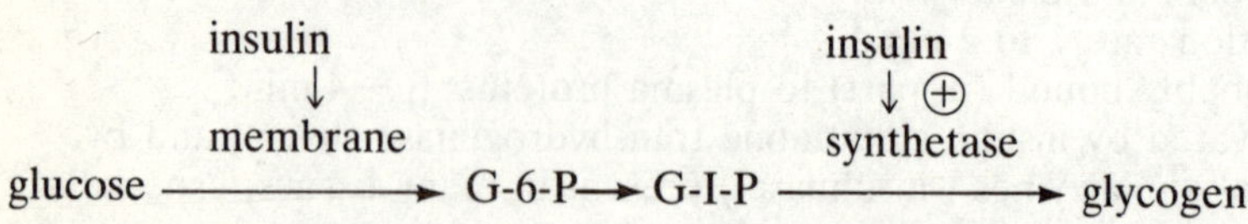

4. ↑ or ↓ glycogenesis (liver)
Insulin decreases glucose output from liver by reducing glycogenolysis and gluconeogenesis. If hypoglycaemia is induced, this action of insulin on the liver is antagonised by adrenaline and glucagon.

insulin →
- ↓ glycogenesis (immediate effect) — due to ↑ glycogenolysis in response to rapid oxidation of glucose mainly by extrahepatic cells.
- ↑ glycogenesis (next few hours) — due to slower oxidation of glucose by extrahepatic cells as energy needs are satisfied and glucose is absorbed from small intestine. Glucokinase also increases in liver.

Liver cells are permeable to glucose and transport (diffusion) of glucose is not affected by insulin.

* Embden-Meyerhof pathway

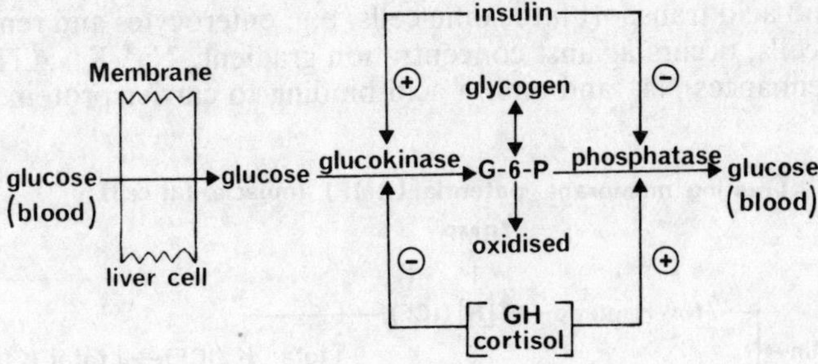

Phosphatase (not present in muscle) dephosphorylates G-6-P and glucose diffuses through membrane into blood.

5. ↑ lipogenesis (fat cells) (See Figures 7-5a, 7-5b)

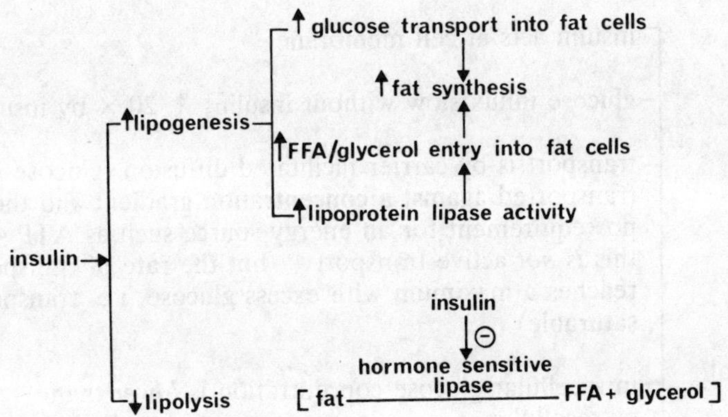

↑ glucose oxidation ⇄ ↓ fat oxidation

6. ↑ protein synthesis

insulin → ↓ protein catabolism (↑ glucose oxidation, 'protein-sparing effect')
insulin → ↑ aminoacid transport into cells
insulin → ↑ m RNA-ribosome activity

Amino acid transport into some cells, e.g. enterocytes and renal tubule cells, occurs against concentration gradient. Na^+-K^+ ATPase pump enhances Na^+ and amino acid binding to carrier protein.

7.

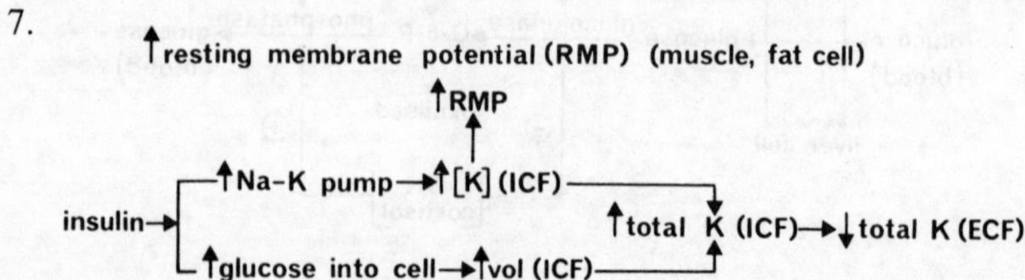

GLUCOSE TRANSPORT ACROSS CELL MEMBRANE (E.G. MUSCLE CELL)

Special points

- insulin acts at cell membrane
- glucose influx slow without insulin; ↑ 20 × by insulin
- transport is by carrier facilitated diffusion (glucose is not transported against a concentration gradient and there is no requirement for an energy source such as ATP — i.e. this is *not* active transport — but the rate of transport reaches a maximum with excess glucose, i.e. transport is saturable)
- intracellular glucose concentration is *lower* than extracellular
- insulin has no effect on glucose transport
  - brain
  - renal tubules
  - intestinal mucosa
  - liver
  - RBC

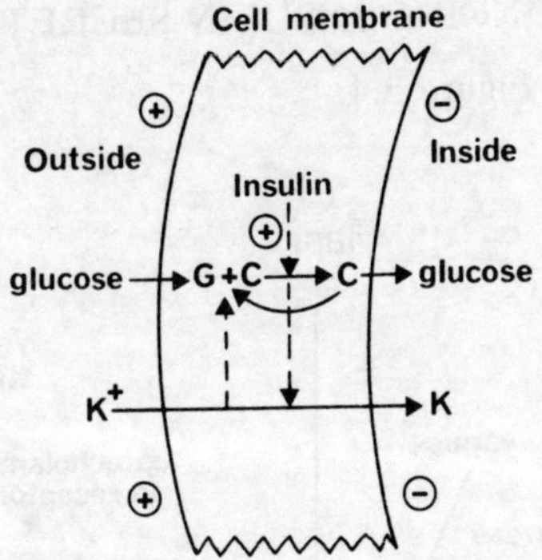

Fig. 8.3 Effect of insulin on glucose transport

Key: C = carrier protein

Glucose entry into cells is enhanced by hypoxia and muscle work.

INSULIN ANTAGONISTS

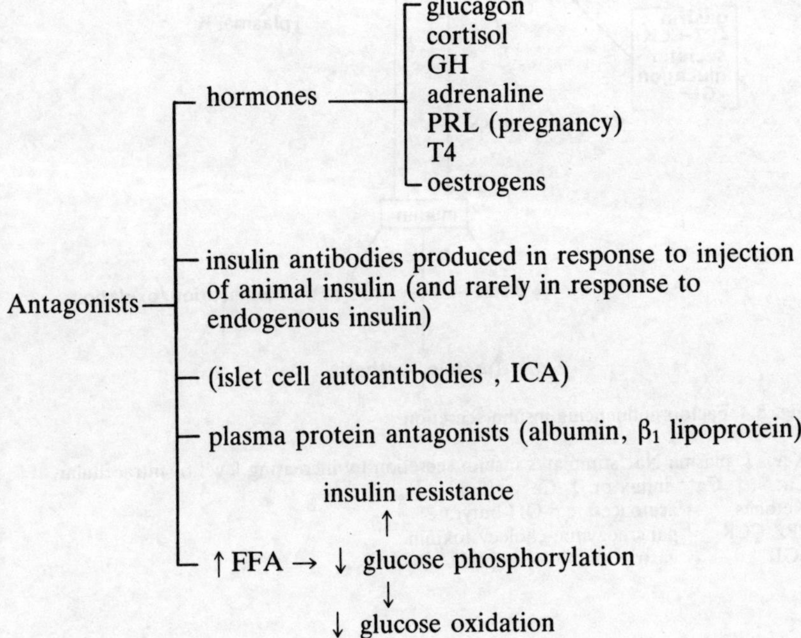

Insulin resistance in juvenile-onset diabetes is due to defective receptor-binding of insulin in addition to the action of antagonists.

FACTORS INFLUENCING INSULIN SECRETION

These are shown in Figure 8.4.

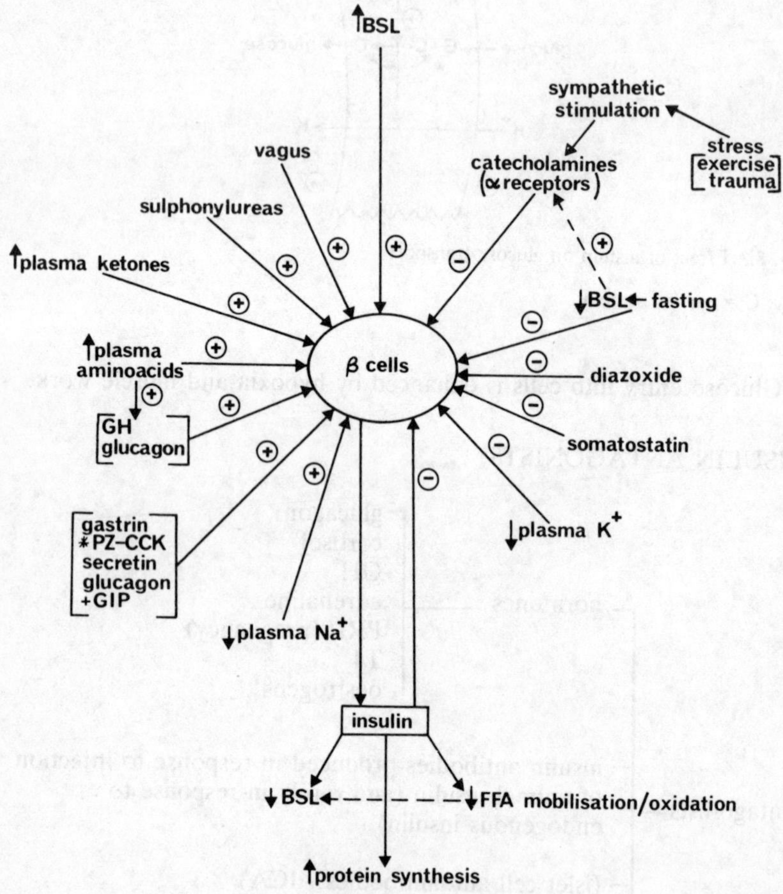

Fig. 8.4 Factors influencing insulin secretion

Key: ↓ plasma Na^+ stimulates insulin secretion by increasing level of intracellular Ca^{++} (↑ Ca^{++} influx or ↓ Ca^{++} efflux)
Ketones = acetoacetate/β-OH-butyrate
*PZ-CCK = pancreozymin-cholecystokinin
+GIP = gastric inhibitory peptide

Adrenaline acting on α receptors inhibits, and on β receptors stimulates, insulin secretion; the former action normally predominates. Glucose activates cyclic AMP system in β cells $\xrightarrow[\text{influx}]{Ca^{++}}$ insulin secretion.

Enteric hormones, especially GIP, potentiate response of β-cells to glucose stimulation.

↑ BSL stimulates glucoreceptors in hypothalamus → vagal excitation → ↑ insulin secretion.

Somatostatin release is stimulated by glucose, amino acids, enteric hormones and cAMP; and is inhibited by sympathetic stimulation and noradrenaline.

EFFECTS OF ADRENALINE AND INSULIN ON BLOOD GLUCOSE CONCENTRATION

These are shown in Figure 8.5.

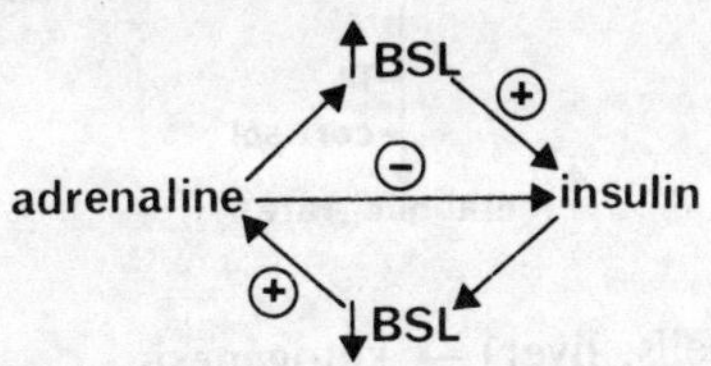

Fig. 8.5 Effects of adrenaline and insulin on blood glucose concentration

GLUCAGON

Polypeptide, 29 amino acids, MW 3485; species differences exist. Secreted by α cells; $t_{\frac{1}{2}}$ plasma <10 min. Enteroglucagon or glucagon-like immunoreactive factor (GLI) cross-reacts with anti-glucagon antibodies in radioimmunoassay systems, and shares some amino acid sequences with glucagon. Glucagon, secretin, vasoactive inhibitory peptide (VIP), and gastric inhibitory peptide (GIP) have some homologous amino acid sequences. Glucagon concentration in plasma: 50 to 150 ng/l (15 to 44 pmol/l).

Glucagon is rapidly degraded in tissues (especially liver and kidney) and by kallikrein in plasma. Some is filtered and reabsorbed through proximal tubules (where it is also metabolised).

Analogues of glucagon have been synthesised including antagonists of glucagon. Glucagon exists as trimers in secretory granules in α-cells; and many 'big' glucagons are present in blood.

Actions of glucagon

1. ↑ glycogenolysis (liver — more active than adrenaline) → ↑ BSL

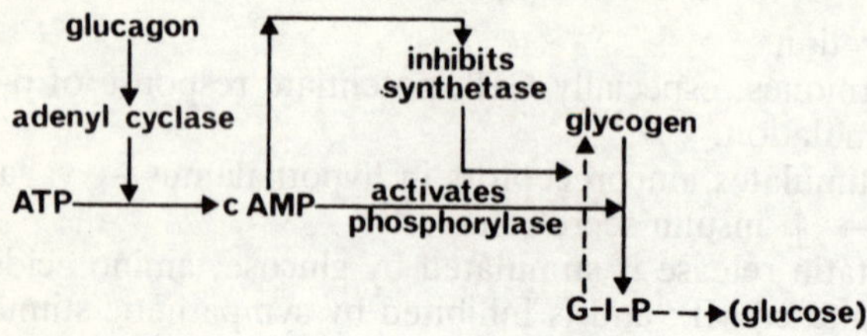

2. ↑ gluconeogenesis

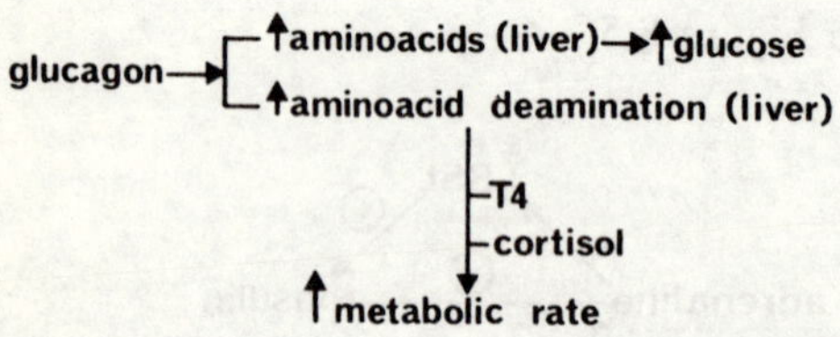

3. ↑ lipolysis (fat cells, liver) → ketogenesis
4. ↑ myocardial contractility and heart rate via adenyl cyclase-cyclic AMP mechanism (in pharmacological doses)

Factors influencing glucagon secretion
These are shown in Figure 8.6.

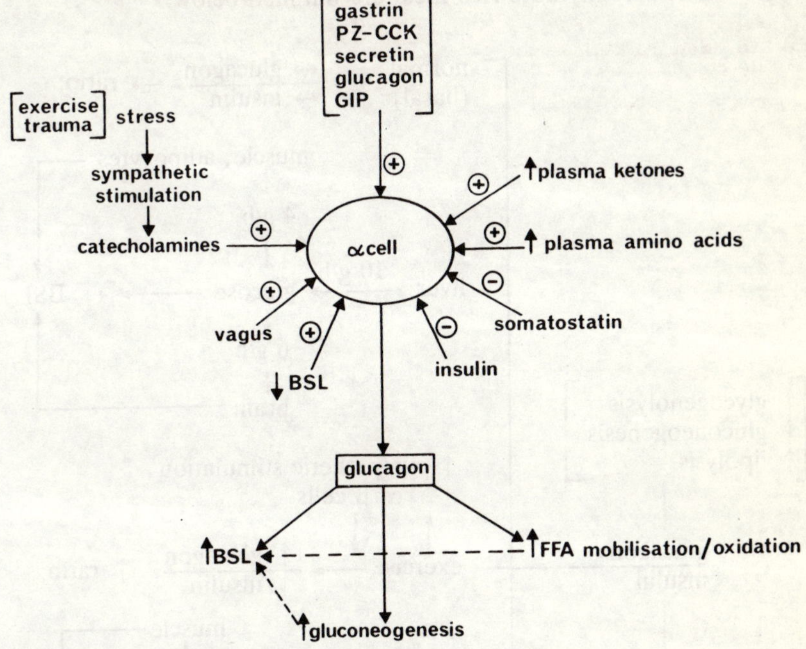

Fig. 8.6 Factors influencing glucagon secretion

Effects of hormones on blood glucose concentration
These are shown in Figure 8.7.

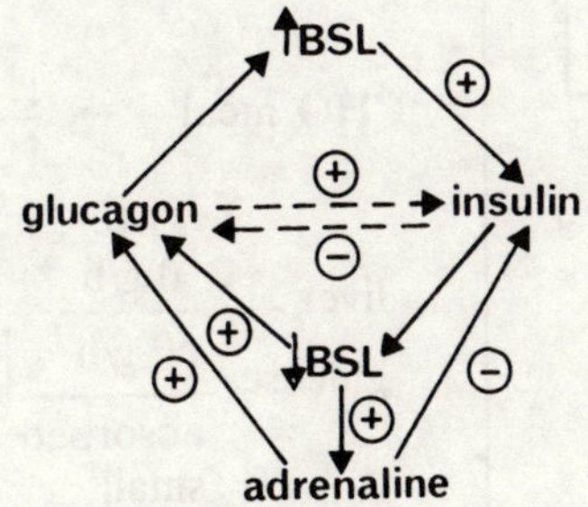

Fig. 8.7 Effects of hormones on blood glucose concentration

Glucagon probably stimulates adrenaline release. Cortisol, GH and T4 → ↑ BSL.

Glucagon: insulin ratio (α/β secretion) and glucose distribution
The glucagon: insulin ratios (and consequent effects on blood glucose level) in normal (basal) state, during exercise and following ingestion of carbohydrate rich meal are outlined below:

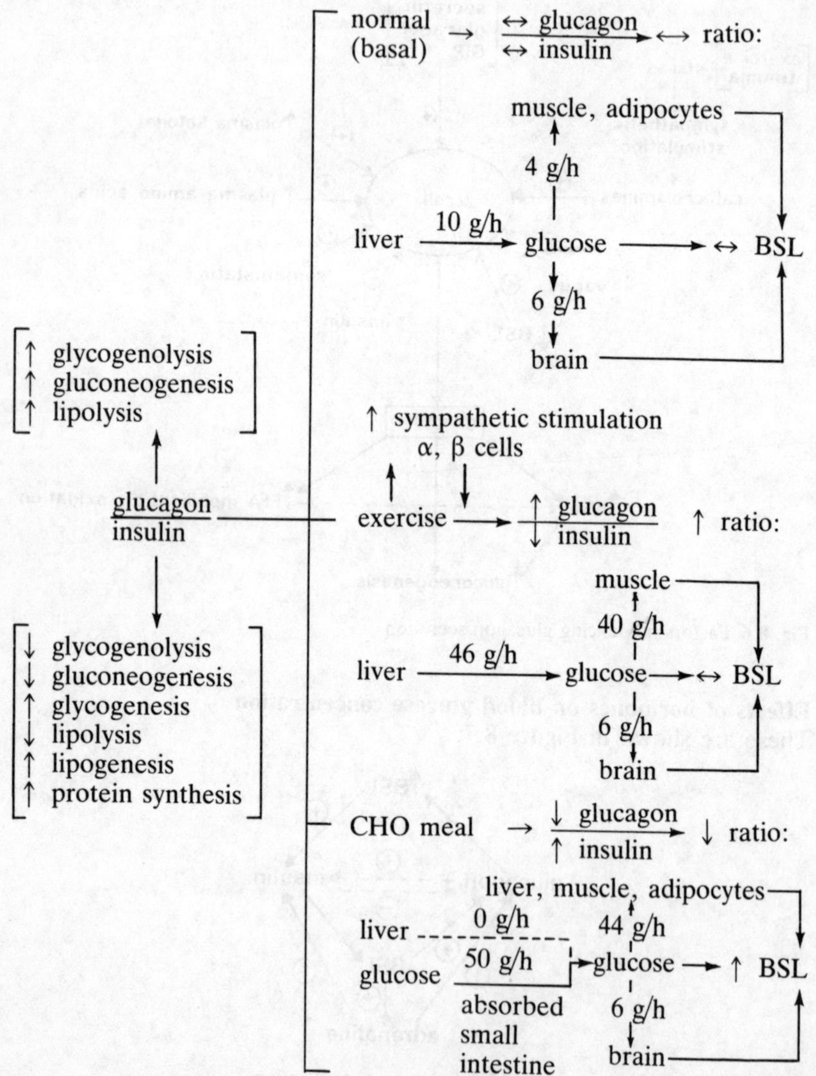

Note: cerebral blood flow is autoregulated (mainly due to PCO_2) and is reasonably steady during course of a day; hence, glucose supply/uptake by brain is held steady.

Artificial pancreas and transplantation studies
The artificial pancreas consists of a large assembly of equipment for analysing glucose concentration used in association with a computer, reservoirs of insulin and glucose, and suitable pumps.
Microminiature forms of the artificial pancreas (about the size of a cardiac pacemaker) which can be implanted are being investigated.

Transplantation of whole pancreas (especially fetal pancreas) and transplantation of islet cells in diabetes is also being studied. A major problem is that of immunorejection.

PHYSIOLOGICAL EFFECTS OF INSUFFICIENT INSULIN SECRETION

a. Metabolic effects
The metabolic effects are outlined in Figure 8.8.

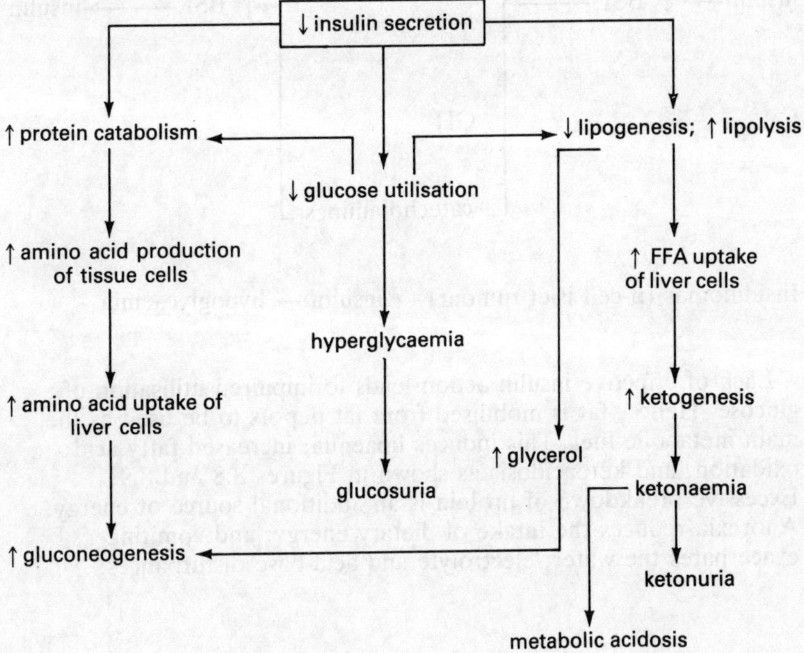

Fig. 8.8 Metabolic effects of insufficient secretion of insulin

Note: ↓ glucose uptake by hypothalamic cells → ↓ inhibition of satiety centre → hyperphagia → ↑ BSL

Insulin and pertinent hormone relationships are as follows:

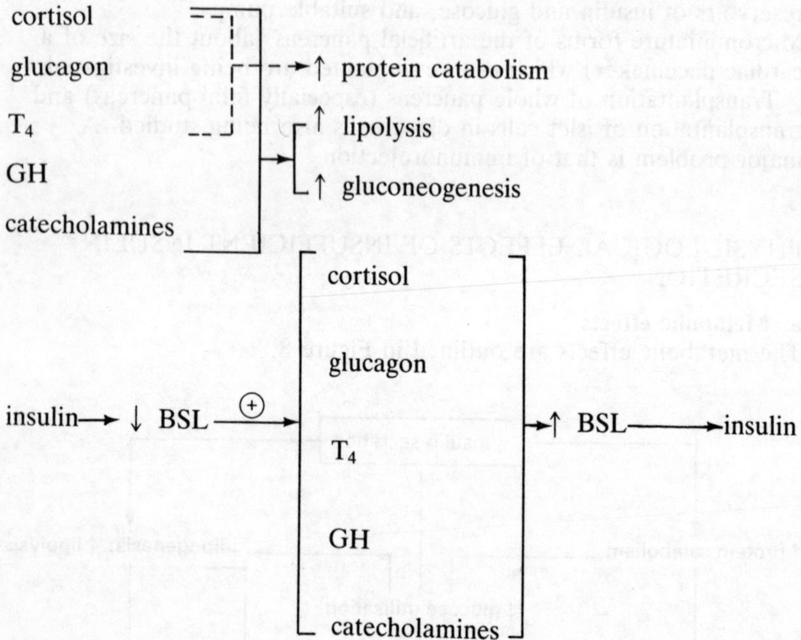

Insulinomas (β-cell islet tumour) → insulin → hypoglycaemia

Lack of effective insulin action leads to impaired utilisation of glucose. Hence, fat is mobilised from fat depots to be used as the main metabolic fuel. This induces lipaemia, increased fatty acid oxidation, and ketoacidosis as shown in Figures 8.8 and 8.9. Excessive breakdown of protein is an additional source of energy. Anorexia reduces the intake of dietary energy; and vomiting exacerbates the water, electrolyte and acid-base disturbances.

b. Other physiological effects of insufficient insulin secretion

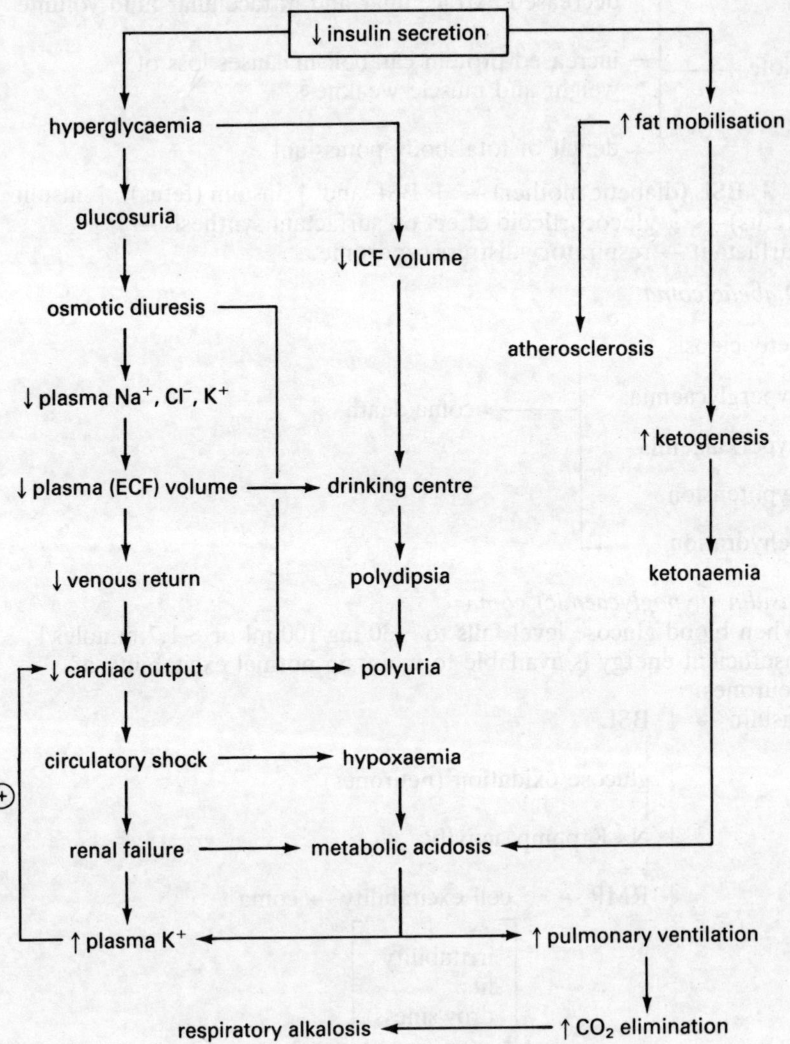

Fig. 8.9 Other physiological effects of insufficient secretion of insulin

Drowsiness, unconsciousness and coma may follow.

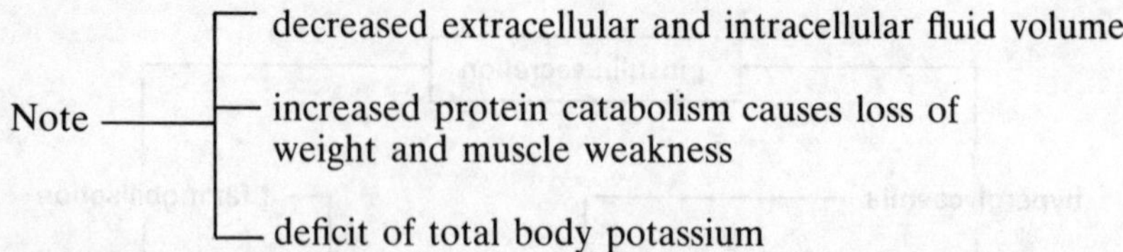

↑ BSL (diabetic mother) → ↑ BSL and ↑ insulin (fetus); ↑ insulin (fetus) → ↓ glucocorticoid effect on surfactant synthesis → ↓ surfactant → respiratory distress syndrome.

Diabetic coma

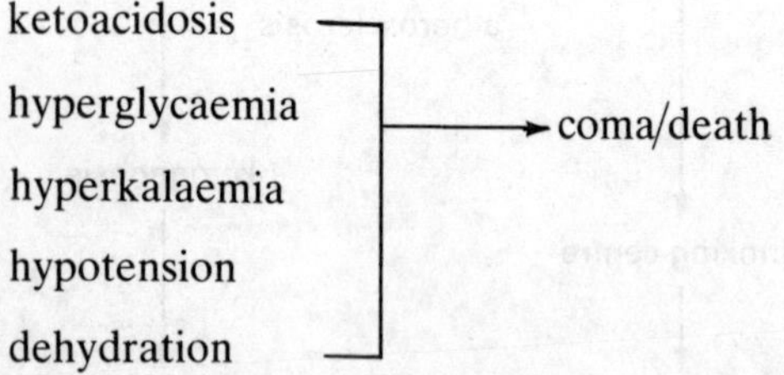

Insulin (hypoglycaemic) coma
When blood glucose level falls to ~30 mg/100 ml or ~1.7 mmoles/l insufficient energy is available to maintain normal excitability of neurones.

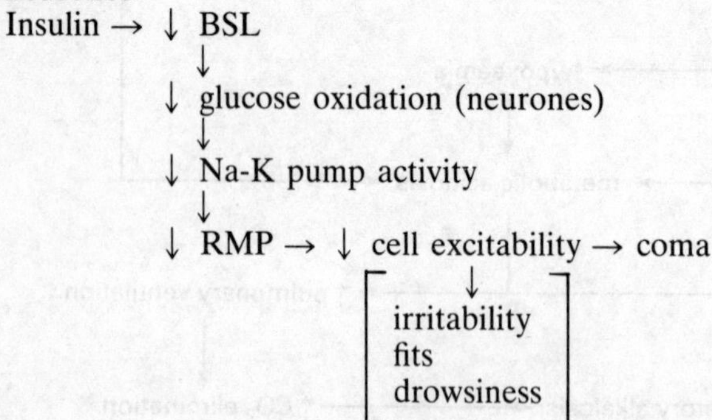

PHYSIOLOGICAL EFFECTS OF EXCESSIVE GLUCAGON SECRETION

These are shown in Figure 8.10; and it should be noted that relative or absolute hyperglucagonaemia is present in clinical diabetes and this contributes to the metabolic disturbances, particularly ketogenesis.

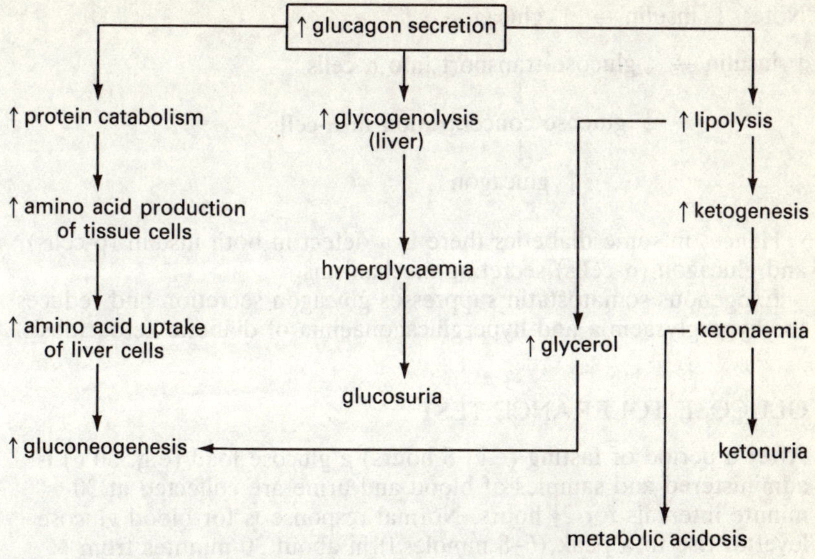

Fig. 8.10 Metabolic effects of excessive secretion of glucagon

Glucagon and pertinent hormone and metabolic relationships are as follows:

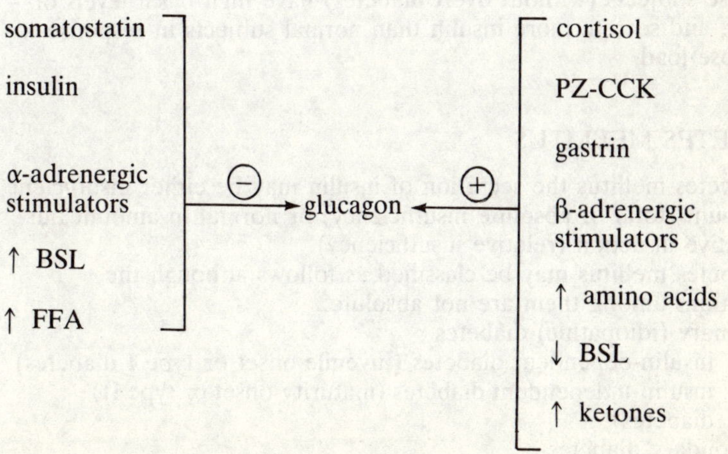

Note: ↓ insulin → ↑ glucagon

↓ insulin → ↓ glucose transport into α-cells
↓
↓ glucose concentration in α-cells
↓
↑ glucagon

Hence, in some diabetics there is a defect in both insulin (β-cells) and glucagon (α-cells) secreting mechanisms.

Exogenous somatostatin suppresses glucagon secretion and reduces the hyperglycaemia and hyperglucagonaemia of diabetic ketoacidosis.

GLUCOSE TOLERANCE TEST

After a period of fasting (e.g. 8 hours) a glucose load (e.g. 50 g) is administered and samples of blood and urine are collected at 30-minute intervals for 2½ hours. Normal response is for blood glucose level to rise to a peak (6–8 mmoles/l) in about 30 minutes from initial fasting level (4.5 mmoles/l) as glucose is absorbed at a faster rate than it is utilised; at about 2 hours the blood glucose level returns to resting (or below resting) value as glucose is utilised by tissue cells, in response to insulin secretion. Glucosuria does not occur as T_m glucose is not exceeded (provided renal threshold is normal).

In diabetes mellitus, the fasting blood glucose level may be raised, blood glucose level peaks at a higher level (e.g. 10–12 mmoles/l) and at about 60 minutes, and return to resting value is delayed for 3 hours or so.

Obese subjects (without overt diabetes) have high basal levels of insulin, and secrete more insulin than normal subjects in response to a glucose load.

DIABETES MELLITUS

In diabetes mellitus the secretion of insulin may be either insufficient in amount (total or absolute insufficiency) or normal in amount but ineffective in action (relative insufficiency).

Diabetes mellitus may be classified as follows although the distinctions among them are not absolute:

1. Primary (idiopathic) diabetes
   a. insulin-dependent diabetes (juvenile-onset or type I diabetes)
   b. insulin-independent diabetes (maturity-onset or type II) diabetes)
2. Secondary diabetes
   a. pancreatic diabetes
   b. insulin-antagonist diabetes

3. Iatrogenic diabetes
4. Diabetes secondary to liver disease.

1. Primary diabetes

a. insulin-dependent diabetes

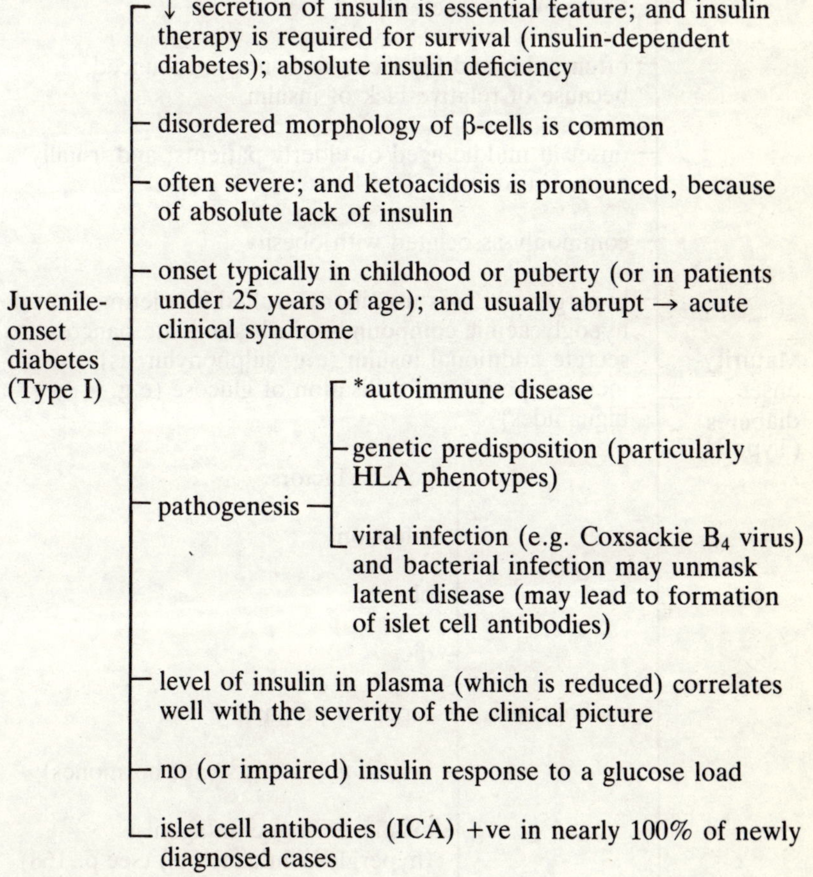

Note: onset of Type I diabetes may be at any age.

* = diabetes often co-exists with autoimmune diseases, e.g. hyperthyroidism, Hashimoto's thyroiditis, Addison's disease, pernicious anaemia; islet-cell antibodies are present in serum of nearly 100% of patients with insulin-dependent diabetes at diagnosis

b. insulin-independent diabetes

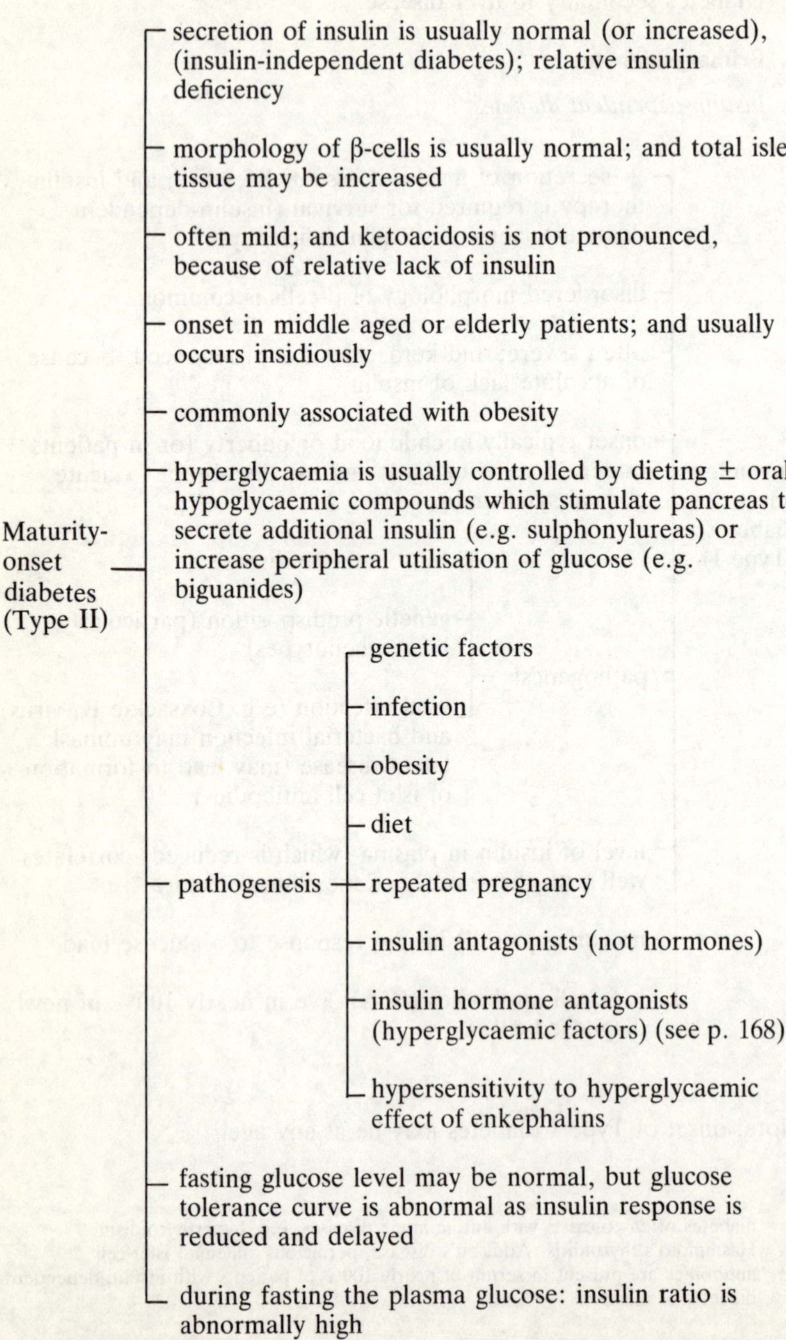

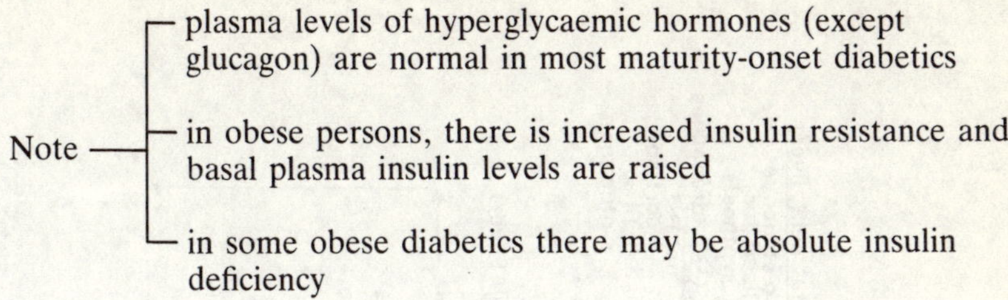

2. Secondary diabetes

a. pancreatic diabetes
Disease processes (e.g. haemochromatosis, pancreatitis) destroy β-cells → ↓ insulin secretion.

There is absolute insulin deficiency (however, some patients can be controlled with sulphonylureas).

b. insulin-antagonist diabetes (endocrine-dependent diabetes)
Hormones (e.g. cortisol, glucagon, thyroxine, growth hormone, adrenaline) are insulin antagonists; hence, the frequent association of diabetes mellitus (or abnormal glucose tolerance curve) with Cushing's syndrome, hyperthyroidism, acromegaly, phaeochromocytoma, pregnancy. In these conditions, hyperglycaemic hormones unmask the underlying diabetic trait probably by placing additional stress on the β-cells.

In endocrine-dependent diabetes there is relative insulin deficiency, i.e. total secretion of insulin is normal (or increased), but effective insulin action is reduced.

3. Iatrogenic diabetes
Cortisol or thiazide diuretics may precipitate diabetes in those with a genetic predisposition.

4. Diabetes secondary to liver disease
Cirrhosis and hepatitis may lead to defects in storage or breakdown of glycogen by liver cells in response to normal secretion of insulin.

Pathophysiology of diabetes mellitus
See Figure 8.11.

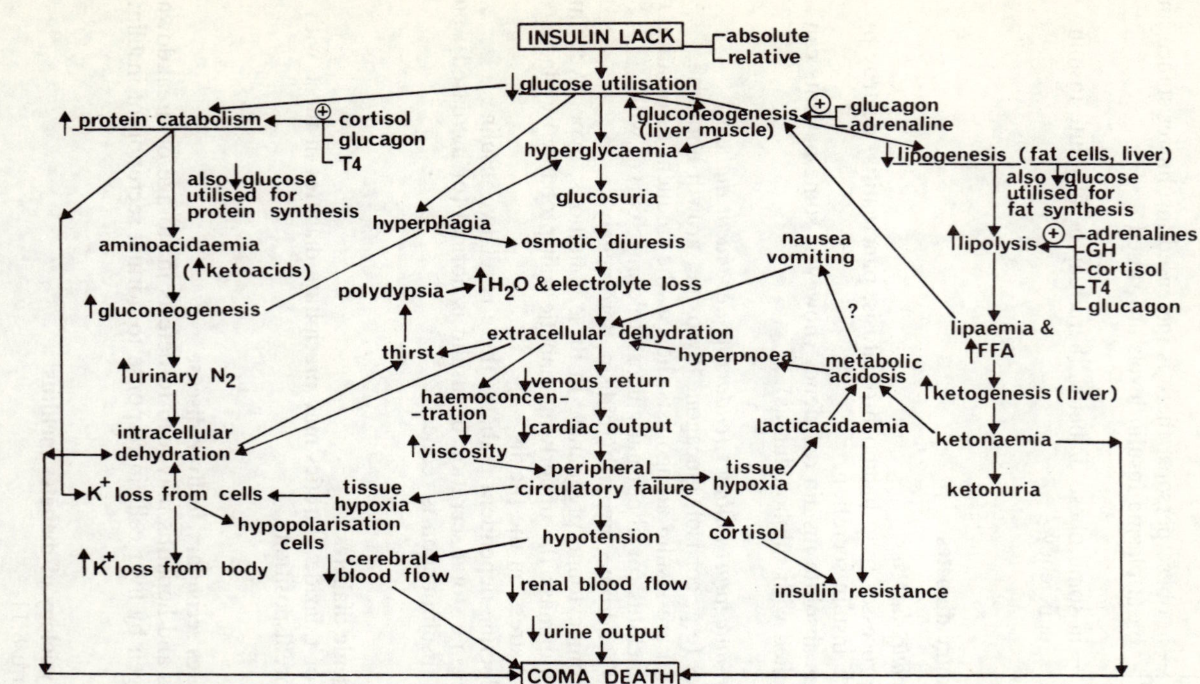

Fig. 8.11 Pathophysiology of diabetes mellitus

Exogenous somatostatin suppresses glucagon secretion and reduces the hyperglycaemia and hyperglucagonaemia of diabetic ketoacidosis. It also reduces glucose absorption from the small intestine. Somatostatin also decreases insulin secretion.

FURTHER READING

Albisser, W. D., Leibel, B. S., Zinman, B., Murray, F. T., Zingg, W., Botz, C. K., Denoga, A., Marliss, E. B. (1977) Studies with an artificial endocrine pancreas. *Archives of Internal Medicine*, **137**, 639.

Baba, S., Saki, H., Imapura, Y. (eds) (1979) *Proinsulin, Insulin, C-peptide*. Amsterdam: Excerpta Medica.

Barseghian, G. R., Leoine, R., Epps, P. (1982) Direct effect of cortisol and cortisone on insulin and glucagon secretion. *Endocrinology*, **111**, 1648.

Bondy, P. K., Rosenberg, L. E. (1980) *Metabolic Control and Disease*. London: Saunders.

Brown, M., Rivier, J., Vale, W. (1976) Somatostatin analogues with selected biologic activities. *Metabolism*, **25** (Suppl.), 1501.

Foa, P. P., Bajaj, J. S. (eds) (1978) *Glucagon: its role in physiology and clinical medicine*. New York: Springer Verlag.

Ganong, W. F., Martini, L. (eds) (1978) *Frontiers in Neuroendocrinology*, vol. V. New York: Raven.

Halban, P. A., Wollheim, C. B., Blondel, B., Meda, P., Niesor, E. N., Mintz, D. H. (1982) The possible importance of contact between pancreatic islet cells for the control of insulin release. *Endocrinology*, **111**, 86.

Narimiya, M., Yamada, H., Matsuba, I., Ikeda, Y., Tanese, T., Abe, M. (1982) The effects of hypoxia on insulin and glucagon secretion in the perfused pancreas of the rat. *Endocrinology*, **111**, 1010.

Smith, P. H., Porte, D. Jr. (1976) Neuropharmacology of the pancreatic islets. *Annual Review of Pharmacology and Toxicology*, 269–285.

Tannenbaum, G. S., Ling, N., Brazeau, P. (1982) Somatostatin-28 is longer acting and more selective than somatostatin-14 on pituitary and pancreatic hormone release. *Endocrinology*, **111**, 101.

Unger, R. H., Orci, L. (1976) Physiology and pathophysiology of glucagon. *Physiology Reviews*, **56**, 778.

Vallance-Owen, J. (ed.) (1975) *Diabetes*. Lancaster: MTP.

Woods, S. C., Porte, D. Jr. (1974) Neural control of the endocrine pancreas. *Physiological Reviews*, **54**, 596–619.

Multiple choice questions

1. Insulin:

1. stimulates K^+ uptake in muscle and adipocyte cells and, thus, hypopolarisation of their membranes;
2. stimulates Na^+-K^+ ATPase in insulin-sensitive cells;
3. alters intracellular Ca^{++} concentration;
4. inhibits adenylate cyclase and stimulates phosphodiesterase in liver and adipocytes;
5. promotes glycogenesis in liver by induction of glycogen synthetase; parasympathetic stimulation also induces glycogen synthetase;
6. binding sites on insulin-sensitive cells are reduced in obese individuals;
7. increases concentration of phosphodiesterase in liver cells and, thus, reduces intracellular cAMP and liver glycogenolysis.

2. Blood glucose concentration tends to be lowered by:

1. vagal stimulation of β-cells;
2. glucose diffusion through ECF and ICF (i.e. total body water);
3. exercise;
4. fasting of 5-days duration;
5. somatomedins;
6. hunger;
7. dehydration.

3. Blood glucose concentration tends to be raised by:

1. hunger;
2. glucagon;
3. cortisol;
4. growth hormone;
5. sympathetic stimulation of α-cells;
6. glucose absorption from stomach;
7. sympathetic stimulation of liver cells.

4. Glucagon:

1. binds to receptors on the hepatocyte membrane; adrenaline binds to the same receptors;
2. and adrenaline bind to their respective receptors on the hepatocyte membrane and activate membrane-bound adenyl cyclase;
3. is secreted directly into the portal blood; this is one factor enabling it to play a more important role than adrenaline in control of hepatic glycogenolysis;
4. increases formation of cAMP in liver cells and, thus, hepatic glycogenolysis;

5. increases FFA mobilisation from adipocytes; these are metabolised preferentially by muscle, thus conserving glucose for cerebral use.

5. With respect to islet cell hormones:
1. glucagon and enteroglucagon are the same molecule;
2. sympathetic stimulation increases secretion of glucagon and insulin;
3. glucagon is completely absent from blood (in man) when pancreas is completely removed (and glucagon is not administered);
4. somatostatin release is stimulated by raised blood glucose level;
5. somatostatin release is inhibited by sympathetic stimulation.

6. Sympathetic stimulation leads to:
1. inhibition of phosphodiesterase in liver cells and, thus, to increase in cAMP and glycogenolysis;
2. release of adrenaline from adrenal cortex;
3. release of glucagon from α-cells of pancreas;
4. release of insulin from β-cells of pancreas;
5. lipolysis in adipocytes.

7. With respect to islet cells:
1. they function in their secretion of hormones as an integrated mass of cells;
2. β-cells inhibit hormone secretion by α-cells;
3. α-cells stimulate hormone secretion by δ-cells;
4. δ-cells stimulate hormone secretion by α-cells;
5. α-cells inhibit hormone secretion by β-cells.

8. Which *one* is not correct?
1. glucose entry into β-cells stimulates proinsulin synthesis;
2. glucose entry into β-cells stimulates C peptide release;
3. amino acid entry into muscle cells is an active process;
4. analogues of glucagon have been synthesised which inhibit actions of glucagon;
5. 'big' glucagons have been detected in circulating blood;
6. GIP inhibits insulin secretion.

9. Juvenile-onset diabetes has the following characteristics:
1. secretion of insulin may be normal or above normal;
2. ketonaemia and ketoacidosis;
3. basal plasma glucose: insulin ratio is abnormally high;
4. commonly controlled by dietary measures;
5. impaired insulin response to a glucose load.

10. Maturity-onset diabetes has the following characteristics:
1. secretion of insulin is commonly reduced;
2. often coexists with other autoimmune disease, e.g. Hashimoto's thyroiditis;
3. commonly associated with obesity;
4. β-islet cell tissue may be increased;
5. secretion of glucagon is commonly reduced.

11. Decreased secretion of insulin is associated with:
1. decreased utilisation of glucose by brain cells;
2. decreased hepatic gluconeogenesis;
3. decreased ICF volume;
4. decreased protein synthesis;
5. increased protein catabolism;
6. negative nitrogen balance;
7. hyperglycaemia which raises osmolarity of plasma.

12. Insulin lack leads to:
1. decreased synthesis of fat in adipocytes;
2. increased hydrolysis of fat in adipocytes and increased FFA transport in blood;
3. decreased oxidation of FFA in liver cells;
4. increased oxidation of acetoacetate in muscle cells;
5. decreased blood volume;
6. hypotension and decreased GFR.

13. The following occur in response to insulin deficiency:
1. obesity with increased insulin resistance;
2. net movement of K^+ from ICF to ECF;
3. net movement of Mg^{++} from ICF to ECF;
4. increased loss of Mg^{++} in urine;
5. vulvitis or balanitis;
6. increased loss of phosphate in urine.

14. Which of the following statements are correct:
1. the concentration of glucose in arterial and venous blood is the same in insulin-dependent (juvenile-onset) diabetes;
2. renal glycosuria is accompanied by hyperglycaemia, polyuria and polydipsia;
3. alimentary (lag) glycosuria results from an abnormally large amount of glucose being absorbed from the ascending colon;
4. carbohydrate deprivation in a normal person can lead to a diabetic type of glucose tolerance curve with glycosuria;
5. the glucose tolerance curve in a healthy man rises more sharply and to a higher level when he has been on a high fat low carbohydrate diet than on a high carbohydrate low fat diet in the preceding several days;

6. in diabetic ketoacidosis, hyperkalaemia may occur in the presence of a total body deficit of potassium.

15. Insulin differs from growth hormone in that it:
1. stimulates hormone sensitive lipase in fat cells;
2. stimulates lipoprotein lipase in vicinity of fat cells;
3. prevents the activation of prolipase in the pancreas;
4. is produced in greater quantity when the amino acid level of the plasma is increased;
5. increases the transport of amino acids across the cell membranes of muscle and heart;
6. increases glucokinase activity in liver cells;
7. is secreted in response to changes in the plasma glucose level;
8. increases m-RNA/ribosome activity;
9. decreases in the plasma in response to somatostatin.

16. In the diagnosis of diabetes mellitus:
1. plasma glucose measurement gives values approximately 15% higher than those in whole blood;
2. glucose tolerance is progressively impaired with increasing age in a Western community;
3. postprandial glycosuria is a reliable and definitive test;
4. approximately 25% of patients with acromegaly have clinical diabetes;
5. a history of diabetes in one parent indicates that children of the marriage have a high risk (>60%) of developing diabetes;
6. fasting plasma insulin levels may be higher than normal in maturity-onset disease.

17. In diabetic ketoacidosis:
1. acidosis and elevated levels of free fatty acids are important contributors to insulin resistance;
2. the average degree of fluid depletion is 5 or 6 litres;
3. the reagent Ketostix gives positive reactions with acetoacetate and β-OH-butyrate;
4. the most important causative mechanism is a marked reduction in insulin action;
5. free fatty acid levels are characteristically low;
6. patients affected are usually of the juvenile-onset insulin-dependent type.

18. The pancreas secretes:
1. inactive precursors of trypsin, chymotrypsin, lipase and amylase;
2. PZ-CCK;
3. insulin, in response to a raised plasma K^+ level;
4. 1 to 2 mg of insulin/day;

5. glucagon, in response to gut glucagon;
6. insulin, in increased quantity when the plasma pH increases;
7. converting enzyme for angiotensin I;
8. gastrin from δ-cells, as evidenced by immunofluorescence studies;
9. proinsulin.

19. With regard to clinical states of hypoglycaemia:
1. the combination of an elevated fasting plasma insulin level and hypoglycaemia is diagnostic of the presence of an insulinoma;
2. an exaggerated response of plasma insulin to intravenous tolbutamide administration is seen in patients with reactive hypoglycaemia;
3. multiple endocrine adenomatosis, Type I, is characterised by tumours of the pancreas, parathyroids and pituitary;
4. excess alcohol intake may cause hypoglycaemia by causing suppression of gluconeogenesis;
5. large extra-pancreatic tumours cause hypoglycaemia by ectopic production of insulin;
6. glucagon administration causes a sharp fall in the elevated levels of insulin seen after glucose administration.

20. In diabetic ketoacidosis:
1. serum osmolality commonly exceeds 330 mOsm/kg water;
2. serum sodium is usually less than 140 mEq/l;
3. considerable pH depression is due to protein catabolism and amino acid accumulation;
4. hyperventilation is invariable;
5. leucocytosis is common;
6. blood pH is commonly below 7.0.

21. In the therapy of diabetic ketoacidosis:
1. early potassium replacement is inadvisable;
2. rapid correction of acidosis is not without danger;
3. other appropriate treatment assumed, several hours delay in insulin treatment is safe;
4. substantial recovery can be expected within 8 to 10 h;
5. complete restoration of metabolic balance can be expected within 48 h;
6. CSF pH falls as blood pH rises.

22. With respect to the pancreas:
1. insulin (like many other protein hormones) is secreted by emeiocytosis (reverse pinocytosis) and this process requires energy;

2. in proinsulin, the C peptide chain links the A and B peptides into position for bridging by disulphide bonds: when these bonds are ruptured the molecule becomes physiologically inactive;
3. proinsulin when secreted into the blood is converted into insulin by trypsin;
4. glucagon has similar metabolic actions to adrenaline;
5. insulin is so rapidly destroyed in brain cells that it has no effect on glucose metabolism in neurones;
6. insulin acts in the testis to increase fructose utilisation by sperm cells;
7. insulin has a vital regulatory role in the metabolism of glucose, amino acids and FFA.

23. Ketoacidosis:
1. never occurs in non-insulin-dependent diabetes;
2. can be recognised by its clinical features;
3. urgently requires glucose and insulin for correction;
4. urgently requires saline and insulin for correction;
5. may be associated with hypokalaemia;
6. is accompanied by considerable electrolyte deficiencies.

24. Hypoglycaemia:
1. occurs after 3 days of fasting;
2. may be prevented by gluconeogenesis;
3. may be due to hyperinsulinism;
4. may occur early in the evolution of diabetes mellitus;
5. can be due to massive retroperitoneal tumours.

25. Diabetes mellitus in obese adults:
1. is due to lack of insulin in the circulation;
2. is associated with elevated levels of insulin in the circulation;
3. is generally improved by weight reduction;
4. requires insulin therapy for control;
5. may be responsible for vaginitis or balanitis.

26. Young insulin-dependent diabetics:
1. are generally underweight at time of diagnosis;
2. often present with dehydration;
3. in the ketoacidotic state have a substantial mortality rate;
4. may eventually become controllable on oral hypoglycaemic drugs;
5. should avoid exercise.

27. With respect to protein synthesis: which is the odd one out?
1. insulin;
2. prolactin;
3. testosterone;

4. oestradiol;
5. growth hormone;
6. cortisol;
7. progesterone;
8. somatomedin.

28. With respect to gluconeogenesis: which is the odd one out?
1. cortisol;
2. glucagon;
3. adrenaline;
4. insulin;
5. thyroxine.

29. With respect to glucose utilisation: which is the odd one out?
1. liver;
2. muscle (skeletal);
3. heart;
4. brain;
5. fat cells.

30. Cortisol differs from glucagon in that it:
1. increases gluconeogenesis during starvation of 2 or 3 days duration;
2. increases lipolysis in fat cells during starvation of 5 days duration and, as a consequence, ketone body production is increased;
3. increases the plasma glucose level;
4. increases hepatic glucogenesis;
5. activates hepatic phosphorylase;
6. can expand the ECF volume in high dosage due to its mineralocorticoid activity.

31. With respect to carbohydrate metabolism: which is the odd one out?
1. glucagon:
2. adrenaline;
3. thyroxine;
4. insulin;
5. cortisol;
6. vasopressin;
7. growth hormone.

32. With respect to fat metabolism: which is the odd one out?
1. growth hormone;
2. adrenaline;
3. cortisol;
4. insulin;
5. glucagon;
6. sympathetic stimulation of adipose tissue.

33. Which of the following are true statements?
1. in hypopituitarism, administration of TSH for three days is followed by an increase in ^{131}I uptake and circulating T4;
2. in a patient with primary aldosteronism and a serum potassium of 2.5 mEq/l, urinary potassium would probably be less than 20 mEq/day;
3. in primary aldosteronism, administration of α-fluorohydrocortisone 0.1 mg four times daily for seven days causes suppression of urinary aldosterone to less than 5 μg/day;
4. failure of plasma renin activity to increase after 5 days of sodium-free diet and after 2 hours of upright posture is diagnostic of primary aldosteronism;
5. in secondary aldosteronism both plasma renin activity and aldosterone secretion are elevated;
6. the oral glucose tolerance test requires carbohydrate restriction for several days before the test;
7. the oral glucose tolerance test will detect the prediabetic state;
8. ketoacidosis in diabetes is best corrected by small doses of insulin frequently;
9. secondary amenorrhoea often follows rapid weight loss in adolescence;
10. secondary amenorrhoea is sometimes seen following treatment with phenothiazine derivatives.

34. During acute haemorrhage (of 20% of blood volume):
1. hyperglycaemia results from increased secretion of GH, cortisol and catecholamines; and from reduced secretion of insulin due to catecholamines and to vasoconstriction in pancreas;
2. hyperglycaemia facilitates expansion of the ECF volume by an osmotic shift of water from the ICF compartment;
3. plasma angiotensin level is increased, and this stimulates cells in the hypothalamic drinking centre;
4. plasma vasopressin level is increased in response initially to reduced central venous pressure;
5. baroreceptors in the JGA are stimulated, and the renin-angiotensin system is activated;
6. plasma aldosterone level is increased, and this causes hyperkalaemic alkalosis;
7. coronary arteries are dilated, and this is due more to local hypoxia than to catecholamines;
8. urine osmolarity falls below that of plasma (330 mOsm/l), because aldosterone avidly increases Na^+ reabsorption from the renal tubules;
9. urine osmolarity increases above that of plasma

(300 mOsm/l), because hyperkalaemia impairs tubular reabsorption of Na^+ by aldosterone;
10. hypokalaemic alkalosis occurs, because aldosterone increases K^+-Na^+ exchange in the distal tubules.

Answers

1. 2, 3, 4, 5, 6, 7
2. 1, 2, 3, 4, 5
3. 1, 2, 3, 4, 5, 7
4. 2, 3, 4, 5
5. 4, 5
6. 1, 3, 5
7. 1, 2, 3
8. 6
9. 2, 3, 5
10. 3, 4
11. 3, 4, 5, 6, 7
12. 1, 2, 4, 5, 6
13. 2, 3, 4, 5, 6
14. 4, 5, 6
15. 2, 6
16. 1, 2, 4, 6
17. 1, 2, 4, 6
18. 1, 3, 4, 5, 8, 9
19. 1, 3, 4
20. 1, 2, 5
21. 1, 2, 3, 4, 6
22. 1, 2, 4, 7
23. 2, 4, 5, 6
24. 1, 2, 3, 4, 5
25. 2, 3, 5
26. 1, 2, 3
27. 6
28. 4
29. 4
30. 5, 6
31. 4
32. 4
33. 1, 5, 7, 8, 9, 10
34. 1, 2, 3, 4, 5, 7

---

9. Gastrointestinal hormones

The gastrointestinal tract has at least 13 different types of endocrine cells which secrete a number of hormones. While these hormones traditionally control gut functions their actions extend into diverse physiological (and pharmacological) processes. Their physiological significance is difficult to establish, and many of the functions ascribed to them are pharmacological and, hence, obfuscate their role in physiological control systems.

Gastrin is the most firmly established hormone, i.e. the administration of pure gastrin produces the physiological responses at plasma levels found to occur after the normal physiological stimulus (usually a meal) for that response. The other hormones are putative or candidate hormones.

Gut hormones can be detected in plasma by radioimmunoassay, and the cells secreting them can be localised by immunocytochemical techniques.

Gastrin (17 amino acids) and pancreozymin-cholecystokinin (PZ-CCK or CCK) (33 amino acids) have amino acid sequence similarities in their structure; and secretin (27 amino acids), glucagon (29 amino acids), gastric inhibitory peptide (GIP) (43 amino acids), and vasoactive intestinal peptide (VIP) (28 amino acids) are other structure/activity related hormones.

Other gut hormones include: motilin, chymodenin, bombesin, enteroxyntin and somatostatin.

In some hormones, fragments of the molecule produce responses e.g. for gastrin, the minimal active fragment is the C-terminal tetrapeptide; for PZ-CCK, the minimal active fragment for contraction of the gall bladder is the C-terminal heptapeptide, and for acid secretion is the C-terminal tetrapeptide. In other hormones e.g. secretin and glucagon, the whole molecule is required for activity.

Some hormones have been detected in a number of different forms. Variants of gastrin are G17 ('little gastrin'), G34 ('big gastrin'); and variants of cholecystokinin are CCK39 and CCK8.

Active analogues of hormones include pentagastrin, and OP-CCK (COOH-terminal octapeptide of CCK).

Hormone interactions have been reported e.g., CCK potentiates action of secretin on bicarbonate secretion by pancreas and reduces action of gastrin on acid secretion by parietal cells.

Trophic effects have been ascribed to some hormones, e.g. gastrin stimulates growth of parietal cells (gastrin-secreting tumours induce hyperplasia and hypertrophy of parietal cells, and antrectomy results in atrophy of these cells).

The distribution of gut hormones is summarised in Table 9.1; stimuli releasing hormones in Table 9.2; actions of hormones on secretion in Table 9.3; and actions of hormones on motility in Table 9.4. Further descriptions are included in the text where differences between physiological and pharmacological actions are not always stressed: hence, there is often apparent (and confusing) overlap of their functions.

Table 9.1 Distribution of hormones

	Antrum	Duodenum	Jejunum	Ileum
Gastrin	+	+		
PZ-CCK		+	+	+
Secretin		+	+	+
*Glucagon		+	+	+
VIP	+	+	+	+
GIP		+	+	+
Motilin		+	+	+
Chymodenin		+	+	

+ = main sites of origin
* = enteroglucagon which is structurally different from pancreatic glucagon

VIP is also found in oesphagus, rectum, pancreas and brain. Bombesin, somatostatin and substance P are present in stomach, small and large intestine.

Table 9.2 Stimuli releasing hormones

	*Protein	*Fat	#Carbohydrate	Acid	Distension
Gastrin	+	0	0	−	+
PZ-CCK	+	+	0	0	0
Glucagon		+	+		
Secretin	0	0	0	+	0
VIP				+	
GIP		+	+		

* = digestive products of protein (e.g. peptides and amino acids) and fat (e.g. monoglycerides and fatty acids)
\# = released by hexoses
\+ = stimulates release of hormones
0 = no effect
− = inhibits release

Alkali in duodenum increases secretion of motilin.

Table 9.3 Actions of hormones on secretion

	Gastric HCl	Pancreatic HCO_3^-	Pancreatic enzymes	Bile HCO_3^-	Pancreatic insulin
Gastrin	+ (P)	+	+	+	+
PZ-CCK	+	+(P)	+ (P)	+ (P)	+
Secretin	−	+ (P)	+ (P)	+ (P)	+
Glucagon	−				+
VIP	−	+	+		+
GIP	−				+

(P) = physiological action

VIP (especially), glucagon, GIP and secretin stimulate secretion of intestinal juice.

Table 9.4. Actions of hormones on motility

	Gastric motility	Gastric emptying	Intestinal motility	Gall bladder contraction
Gastrin	+	+	+	+
PZ-CCK	+	− (P)	+	+ (P)
Secretin	−	−	−	+
Glucagon	−	−	−	+
Motilin	+	+	+	

\+ = stimulates
− = inhibits
(P) = physiological action

GIP inhibits motility of proximal stomach.

Bombesin stimulates gastrin secretion and motility of small intestine and gall bladder. It probably is neurotransmitter at vagal endings on G cells.

Caerulein (isolated from frog skin) has gastrin and PZ-CCK activities, and is used during cholecystography to contract gall bladder.

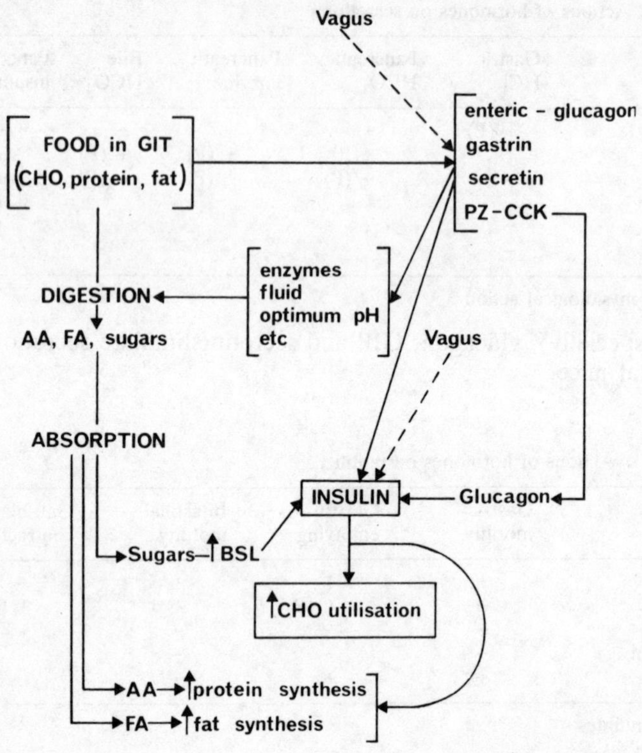

Fig. 9.1 Integration of enteric and pancreatic hormones in digestion and utilisation of foodstuff

Key: AA = amino acids
FA = fatty acids
GIT = gastrointestinal tract
CHO = carbohydrate

Food in GIT → enteric hormones — enzymes etc → digestion
— insulin → utilisation digested products

FURTHER READING

Bloom, S. R. (1977) Gastrointestinal hormones. In Crane, R. K. (ed.) *International Review of Physiology — Gastrointestinal Physiology II*, vol. 12. Baltimore: University Park Press.

Bloom, S. R. (ed.) (1978) *Gut Hormones*. Edinburgh: Churchill Livingstone.

Dockray, G. J. (1979) Evolutionary relationships of the gut hormones. *Federation Proceedings*, **38**, 2295.

Fujimoto, W. Y., Williams, R. H., Ensinck, J. W. (1979) Gastric inhibitory polypeptide, cholecystokinin, and secretin effects on insulin and glucagon secretion by islet cultures. *Proceedings of the Society of Experimental Biology and Medicine*, **160**, 349.

Gregory, R. A. (1974) The gastrointestinal hormones: a review of recent advances. *Journal of Physiology* (*London*), **241**, 1.

Gregory, R. A. (1978) The gastrins: structure and heterogeneity. In Grossman, M., Speranza, V. et al (eds) *Gastrointestinal Hormones and Pathology of the Digestive System*, p. 75. New York: Plenum.

Grossman, M. I. (1979) Neural and hormonal regulation of gastrointestinal function: an overview. *Annual Review of Physiology*, **41**, 27–33.

Grube, D., Forssmann, W. G. (1979) Morphology and function of the entero-endocrine cells. *Hormone and Metabolism Research*, **11**, 589–606.

Hobsley, M. (1981) *Disorders of the Digestive System*. London: Edward Arnold.

Konturek, S. J. (1978) Somatostatin and gastrointestinal secretion and motility. In Grossman, M., Speranza, V. et al (eds) *Gastrointestinal Hormones and Pathology of the Digestive System*, p. 227. New York: Plenum.

Sanford, P. (1981) *Digestive System Physiology*. London: Edward Arnold.

Multiple choice questions

1. Gastrin:
1. is secreted by G cells in the antrum and duodenum;
2. release is stimulated by digestive products of protein in the stomach;
3. stimulates acid secretion by the parietal cells;
4. increases gastric motility;
5. and PZ-CCK have similarities in structure and activity.

2. Pancreozymin-cholecystokinin (PZ-CCK):
1. is secreted by cells in the duodenum;
2. release is stimulated by digestive products of fat in the small intestine;
3. stimulates enzyme secretion by the pancreas;
4. inhibits gastric emptying;
5. potentiates effect of gastrin on acid secretion by the parietal cells.

3. Secretin:
1. is secreted by cells in the duodenum;
2. release is stimulated by an acid medium (e.g. pH 6) in the duodenum;
3. stimulates bicarbonate secretion by the ductular epithelial cells of the pancreas;
4. increases intestinal motility;
5. and motilin have structure/activity relationship.

4. Enteroglucagon:
1. is secreted by cells in the jejunum and colon;
2. release is stimulated by digestive products of fat and by hexoses in the small intestine;
3. stimulates acid secretion by the parietal cells;
4. decreases intestinal motility;
5. secreted by the small intestine and by the pancreas are the same molecule.

5. With respect to gastrointestinal hormones which are correct:
1. vasoactive intestinal peptide (VIP) is present in the stomach, small intestine and colon and causes vasodilatation;
2. the terminal tetrapeptide of gastrin and PZ-CCK are identical;
3. gastric inhibitory peptide (GIP) is abundant throughout the colon;
4. secretin, glucagon, VIP and GIP are a structurally homologous family;
5. 'little gastrin' (G17) is about half as abundant in serum as 'big gastrin' (G34), and is about 5 times more active than 'big gastrin' on acid secretion;

6. serum gastrin activity can be detected by immunoassay at times when there is no known stimulus for G17 release (e.g. during sleep): this activity is due to 'big big gastrin' which apparently is unable to cross capillary walls and thus to stimulate gastrin receptors;
7. 'enterogastrones' inhibit gastric acid secretion and include secretin and VIP.

6. Gastrointestinal hormones influence many gastrointestinal functions. Which of the following statements are true?
1. the primary action of secretin in man is to stimulate pancreatic secretion of water and bicarbonate;
2. in healthy man, gastrin release is usually stimulated by administration of exogenous secretin;
3. cholecystokinin and pancreozymin have distinctly different modes of action and chemical structures;
4. cholecystokinin stimulates pancreatic enzyme secretion;
5. pancreatic glucagon (greatly) decreases volume flow and enzyme output of the stimulated pancreas;
6. gastric inhibitory polypeptide (GIP) occurs in cells in the duodenal mucosa;
7. vasoactive intestinal polypeptide (VIP) weakly stimulates electrolyte and water secretion by the pancreas, and stimulates secretion and muscle contraction in the small intestine.

7. The actions of gastrin include:
1. reduction in gastrointestinal motility;
2. increased output of pancreatic enzymes;
3. decreased output of bile;
4. increased output of pepsinogen;
5. decrease in tone of lower oesophageal sphincter;
6. increased output of hydrogen ion by the parietal cell.

8. Secretin differs from PZ-CCK in that it:
1. is a peptide;
2. decreases production of gastric HCl;
3. decreases gastric motility;
4. decreases motility of small intestine;
5. increases insulin production;
6. increases markedly pancreatic bicarbonate secretion;
7. increases Brunner's glands secretion;
8. is an 'enterogastrone'.

9. With regard to the gastrointestinal hormones:
1. gastrin is the most potent known stimulus to gastric acid secretion;

2. fasting serum gastrin levels are significantly lowered by somatostatin (GIF);
3. gastrin levels are characteristically suppressed in the Zollinger-Ellison syndrome;
4. gastrin stimulates intrinsic factor secretion;
5. secretin inhibits pancreatic bicarbonate secretion;
6. the greater release of insulin seen following oral rather than intravenous glucose administration is due to the release of gastrointestinal hormones;
7. serotonin stimulates large bowel motility.

10. Gastrin differs from PZ-CCK in that it:
1. is a more potent stimulant of HCl production by the gastric parietal cells;
2. increases gastric motility;
3. increases insulin production;
4. increases pepsin production by the gastric chief cells;
5. is a peptide with a terminal core of 5 amino acids which are shared by PZ-CCK;
6. is produced by G cells mainly in the pyloric antrum;
7. is produced in greater quantity in achlorhydria with pernicious anaemia;
8. is produced in greater quantity following vagotomy.

11. Gastrin differs from PZ-CCK in that it:
1. can be produced by △ cells in the pancreas;
2. is a less potent stimulant of enzyme secretion by the pancreas;
3. is produced in greater quantity following gastrojejunostomy, when the antrum is displaced surgically from contact with the acid chyme;
4. is produced in greater quantities in the Zollinger-Ellison syndrome;
5. is a less potent stimulant of the gall bladder smooth muscle;
6. increases production of pancreatic glucagon;
7. is secreted in response to protein products;
8. increases motility of the small intestine.

12. Which is the odd one out?
1. insulin;
2. glucagon;
3. secretin;
4. PZ-CCK;
5. LATS;
6. TSH;
7. oxytocin;
8. vasopressin.

Answers

1. 1, 2, 3, 4, 5
2. 1, 2, 3, 4
3. 1, 3
4. 1, 2, 4
5. 1, 2, 4, 5, 6, 7
6. 1, 4, 5, 6, 7
7. 2, 4, 6
8. 2, 3, 4, 6, 8
9. 1, 2, 4, 6, 7
10. 1, 6, 7, 8
11. 1, 2, 3, 4, 5
12. 5

10. Parathyroid glands

CALCIUM METABOLISM AND PHYSIOLOGY OF BONE

The parathyroid glands and calcium metabolism
Plasma calcium is maintained within narrow limits by homeostatic mechanisms. The ionised calcium is the active plasma fraction that regulates parathyroid function by negative feedback (see Fig. 10.3), and may act at steps (1), (2) and (3) in PT cell as shown in Figure 10.2.

Plasma calcium fractions

Total (*10 mg/100 ml) —
- ionised 4.5 mg
- complexed 1 mg (to citrate, phosphate, bicarbonate)
- protein bound 4.5 mg

Ca^{++} + $Protein^{-} \rightleftharpoons$ Ca — Protein complex

$$\frac{[Ca^{++}]\,[Prot]}{[Ca\ Prot]} = K \text{ (dissociation constant)}$$

Parathyroid hormone (PTH) and vitamin D are of prime importance in calcium homeostasis. Calcitonin has no definite physiological role in man. Daily calcium turnover is shown in Figure 10.1.

The parathyroid glands
Total weight of the four glands is 120 to 140 mg; principal cell type is 'chief' cell which secretes PTH. Pale oxyphil cells are also present and large groups of fat cells are scattered throughout the glands.

* 5 mEq/l or 2.5 mM/1

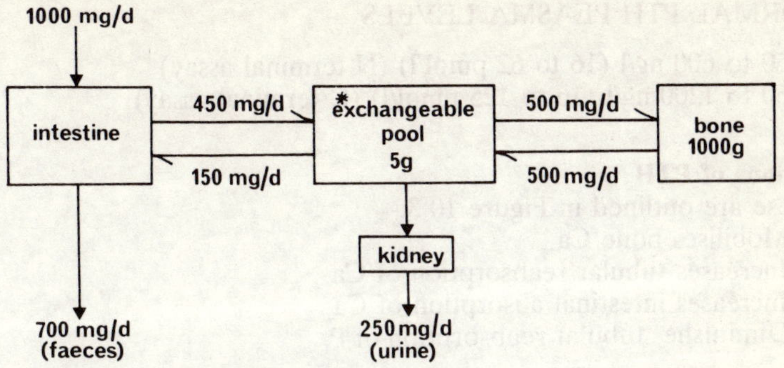

Fig. 10.1 Calcium metabolism

Key: * = in bone surfaces, soft tissues, and ECF
About 50 mg/d are lost in sweat

Parathyroidectomy → hypocalcaemia (with tetany)
Injection of PTH → hypercalcaemia
Excess PTH → hypercalcaemia
secretion

Parathyroid hormone (PTH)
84 amino acid single chain polypeptide, MW 9500, formed by cleavage from pro-PTH (90 amino acids) prior to release from the gland. Activity of PTH resides in the 1 to 34 amino terminal fragment. Very little PTH is stored in the glands; $t_{\frac{1}{2}}$ in plasma is 20 minutes. Small amounts of pro-PTH and PTH fragments may be secreted with PTH.

PTH is metabolised in the Kupffer cells of the liver (and possibly other tissues) and fragments of PTH are produced which are reactive in the radioimmunoassay for PTH (and possibly have some biological activity) (see Fig. 10.2). The kidney excretes PTH and PTH fragments.

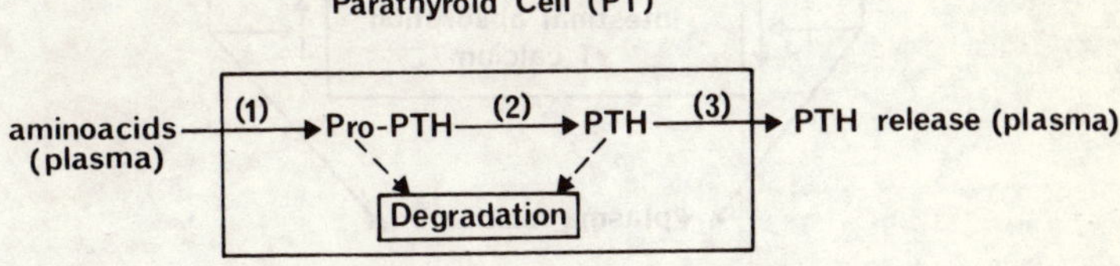

Fig. 10.2 Synthesis and release of PTH

NORMAL PTH PLASMA LEVELS

<150 to 600 ng/l (16 to 62 pmol/l) (N terminal assay)
<150 to 1200 ng/l (16 to 125 pmol/l) (C terminal assay).

Actions of PTH

These are outlined in Figure 10.3.

1. Mobilises bone Ca
2. Increases tubular reabsorption of Ca
3. Increases intestinal absorption of Ca
4. Diminishes tubular reabsorption of P_i

1. PTH and mobilisation of bone calcium

Bone contains about 1 kg calcium of which 5 g is exchangeable; ECF contains about 1 g (25 mmol) calcium (see Fig. 10.1).

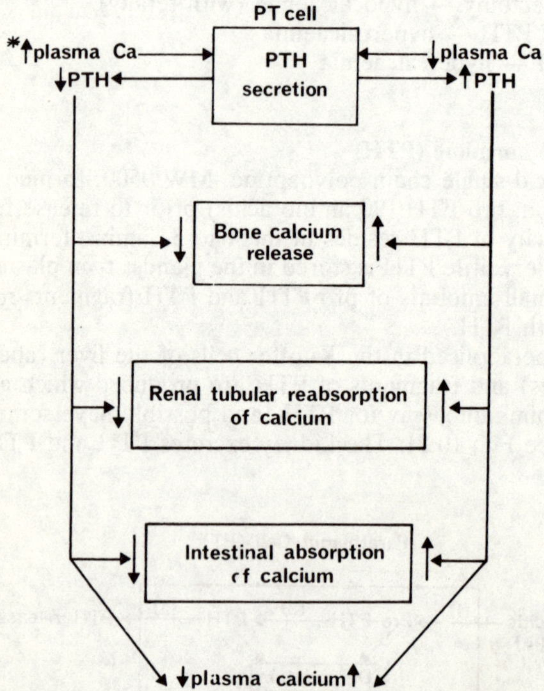

Fig. 10.3 Actions of PTH

Key: *changes in plasma Mg^{++} act in same direction as those in plasma Ca^{++}

PTH has two main actions on bone calcium:

a. rapid mobilisation, within 10 minutes — mechanism unknown (probably too fast to be mediated by resorption by bone cells)

b. long continued mobilisation, mediated by increased osteoclastic and osteocytic activity leading to osteolysis (bone resorption), PTH activates osteoclasts (DNA synthesis) and osteocytes in 1/2 to 2 h ; and inhibits osteoblastic activity.

Bone cell response to PTH

a. PTH increases intracellular Ca → activation cell enzymes

b. PTH acts via adenylate cyclase

Note: The *earliest* effect of PTH on plasma Ca is a transient fall. Vitamin D is necessary for normal bone cell function.

c. subsequently (12 h +) PTH stimulates production of collagenase by osteoclasts → destruction bone collagen (→ release of free and peptide hydroxyproline ,which increase in plasma and in urine).

2. PTH and renal calcium excretion

PTH decreases urine Ca before plasma Ca increases.
PTH increases % of filtered Ca that is reabsorbed.

Filtered Ca = 7 to 10 g/24 h
Urine Ca = 100 to 400 mg/24 h.

PTH acts on distal (not proximal) tubular absorption of Ca^{++}; mechanism is unknown and no role for cyclic AMP has been defined. Rapid physiological adjustments in plasma Ca are brought about through the effects of PTH on tubular reabsorption of Ca^{++}.

3. PTH and intestinal calcium absorption

Dietary calcium = 0.5 to 1.0 g (approx.) daily.
Absorption occurs in three steps:

a. across luminal surface of intestinal cell by facilitated diffusion or active transport (increased by vitamin D)

b. transport within the intestinal cell in association with a Ca-binding protein; formation of Ca-binding protein is increased by vitamin D

c. Na^+-dependent active transport across serosal surface of the intestinal cell (most active in duodenum; increased by vitamin D).

Dietary Pi has probably no direct effect on Ca absorption.

Effects of low dietary calcium
Low dietary Ca intake results in increased efficiency of intestinal Ca absorption by:
1. Increased affinity of Ca-binding protein
2. Increased PTH secretion which leads to
3. Increased conversion of 25(OH)CC to $1,25(OH)_2CC$ (see Fig. 10.4).

PTH increases Ca absorption within about 24 h.

4. PTH and renal Pi excretion
PTH inhibits proximal tubular reabsorption of Pi (distal tubules play little part in Pi reabsorption). It does not cause tubular secretion of Pi.

PTH infusion → ↑ renal cell cyclic AMP. Urinary cyclic AMP rises *before* urinary Pi rises.

With PTH excess: plasma Pi falls, despite release of bone Pi, because of increased renal Pi excretion.

VITAMIN D METABOLISM

This is outlined in Figure 10.4.
Vitamin D deficiency (or ↓ activity of vitamin D) → ↓ Ca absorption from small intestine → ↓ plasma Ca → ↓ Ca available to bone → ↓ mineralisation of bone → osteomalacia (adults), rickets (children).

CALCITONIN (CT)

32 amino acid polypeptide hormone secreted by the parafollicular cells ('c' cells) of the thyroid gland: 'c' cells arise from the neural crest and may also be found in the thymus gland. MW 3421; $t_{\frac{1}{2}} \sim 30$ minutes; activity apparently due to 7 of the 9 N-terminal amino acids; degraded in liver and kidney.

Basal plasma CT levels = 0.02–0.04 ng/ml (5.4–10.8 pmol/l).

Raised by calcium infusion and calcium ingestion. Release is also stimulated by gastrin, PZ-CCK and glucagon.

Principal action of CT is to reduce plasma calcium level by inhibiting resorption of bone mineral.

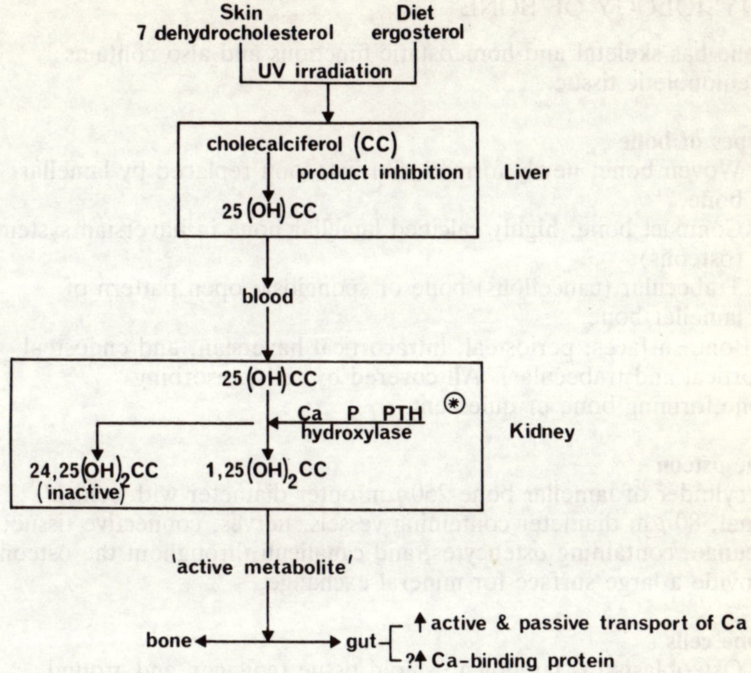

Fig. 10.4 Vitamin D metabolism

Key: ⊛ 1-hydroxylase which converts 25(OH)CC to active 1,25 $(OH)_2CC$ is stimulated by PTH and a low ECF P_i. This enzyme is possibly stimulated by low intracellular Ca^{++}.

25(OH)CC = 25 hydroxycholecalciferol
24,25$(OH)_2$CC = 24,25 dihydroxycholecalciferol
1,25$(OH)_2$CC = 1,25 dihydroxycholecalciferol

Mode of action

1. inhibits osteoclastic resorption of bone
2. ↓ plasma calcium (leads to compensatory ↑ PTH secretion)
3. ↓ intestinal absorption of calcium
4. ↑ urinary Ca and Pi

Physiological importance of CT in man is uncertain; and there are no obvious manifestations of CT deficiency.

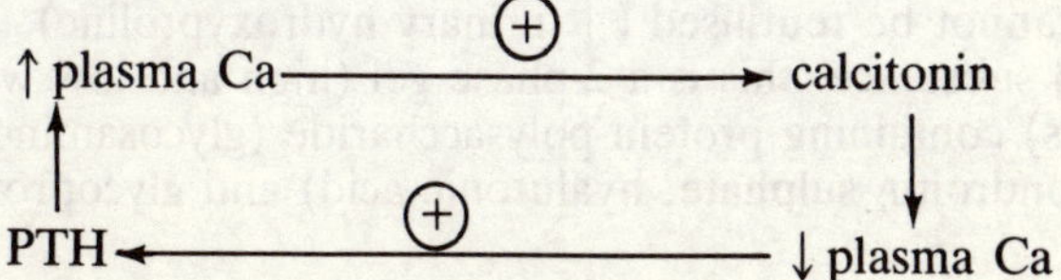

PHYSIOLOGY OF BONE

Bone has skeletal and homeostatic functions and also contains haemopoietic tissue.

Types of bone

1. Woven bone: newly formed, fibrous, soon replaced by lamellar bone.
2. Compact bone: highly calcified lamellar bone in haversian systems (osteons).
3. Trabecular (cancellous) bone or spongiosa: open pattern of lamellar bone.

Bone surfaces: periosteal, intracortical haversian, and endosteal (cortical and trabecular). All covered by cells, resorbing bone,forming bone or quiescent.

The osteon

A cylinder of lamellar bone 250 μm outer diameter with central canal, 80 μm diameter containing vessels, nerves, connective tissue; lacunae, containing osteocytes, and canaliculi throughout the osteon provide a large surface for mineral exchange.

Bone cells

1. Osteoblasts (a) lay down osteoid tissue (collagen and ground substance) in a layer adding 1 μm thickness per day; (b) transfer mineral to osteoid tissue; (c) produce alkaline phosphatase.
2. Osteocytes can form and resorb bone (number = 26 000 per mm^3).
3. Osteoclasts: large multinucleate (2 to 100 nuclei) mobile cells, living from hours to days; brush border resorption by enzymic and acid digestion of bone, forming Howship's lacunae on endosteal surfaces. An osteoblast takes one month to lay down as much bone as an osteoclast resorbs.

Bone matrix

This consists of 90% collagen in ground substance.

1. Collagen formation and breakdown: osteoblasts make and release tropocollagen. This polymerises to triple helix polypeptide chains of collagen. Vitamin C, heparin, growth hormone, parathyroid hormone are factors affecting collagen synthesis. Collagen breakdown (collagenase in osteoclasts) releases hydroxyproline which cannot be reutilised ($\uparrow$ urinary hydroxyproline).
2. Ground substance: this is a 2-phase gel (high and low water contents) containing protein-polysaccharide (glycosaminoglycans, e.g. chondroitin sulphate, hyaluronic acid) and glycoproteins.

Bone mineralisation
75% is initiated enzymatically and rapidly (few days) completed by osteoblasts; Ca^{++} or Pi interacts with 64 nm banded collagen. Sequence is: amorphous → first crystalline form → second crystalline form → hydroxyapatite:

$3\ (Ca_3(PO_4)_2).CaX_2.$
$X_2 = OH_2 + \text{trace of } F_2$

Also present: Na^+ 270 mM/kg, Mg^{++}, K^+, citrate, carbonate. Heavy metals such as lead and nuclear fission products (e.g. strontium) may also be incorporated in bone.

Pyrophosphate inhibits mineralisation in bone. Removal of pyrophosphate may trigger calcification (bone has two pyrophosphatases).

Bone growth and ossification
Appositional (bone laid down on existing surfaces)

Intramembranous ┐
                ├ cease at maturity
Endochondral   ┘

Endochondral growth (→ ↑ length): chondrocytes proliferate, hypertrophy and produce cartilage; the walls of chondrocytes calcify to form spicules; capillary invasion follows; osteoblasts appear and lay down osteoid tissue on the spicules. The osteoid tissue undergoes mineralisation and is then modelled by osteoclasts and osteoblasts.

Bone cells and mineral homeostasis
3 to 5% of bone is undergoing remodelling at a given time.

Exchange of bone calcium depends on bone cell remodelling and metabolic activity.

Formation of bone cells

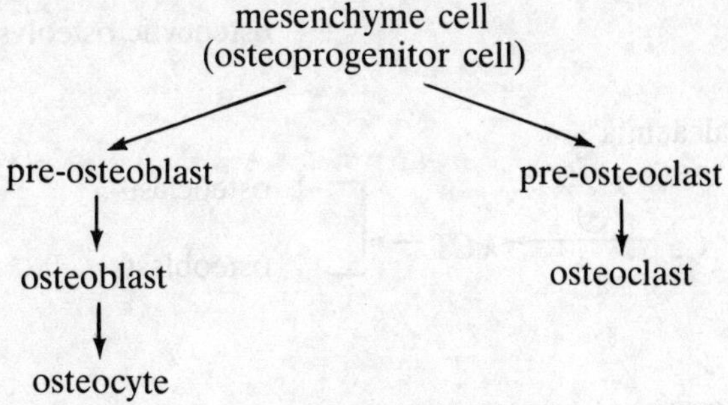

Formation of bone is more active than resorption: bone diameter gradually increases.

The *primary event* is osteoclastic resorption. This is followed by osteoblastic bone formation. Bone is thus remodelled and ageing bone replaced. At a given site, resorption takes 20 days and formation takes 80 days.

Balance favours resorption after 30 to 40 years of age.

The bone remodelling unit gives rise to new bone with a metabolically active life up to 10 years or more. This bone acts as a metabolic unit at its surface and in its lacunar-canalicular system (occupied by osteocytes and their processes).

Balance between resorption and formation is achieved by a coupling mechanism linking osteoblastic and osteoclastic activity.

Agents influencing cellular events on endosteal surfaces

1. Osteoclasts: their formation from osteoprogenitor cells and their activity are
   a. stimulated by PTH, vitamin D, thyroxine and growth hormone;
   b. inhibited by calcitonin, oestrogen and glucocorticoids.
2. Osteoblasts: their formation and their activity are
   a. stimulated by mechanical stress, calcitonin, phosphate and oestrogen;
   b. inhibited by PTH.

Note: Mechanical stress maintains integrity of bone. It sets up piezo-electric forces that may stimulate formation and activity of osteoblasts.

Effects of hypo- and hyper-calcaemia on bone cells

1. Hypocalcaemia

↓ Ca —(+)→ PTH → ↑ osteoclasts; ↓ osteoblasts; ↑ osteocytic osteolysis

2. Hypercalcaemia

↑ Ca —(+)→ ★CT → ↓ osteoclasts; ↑ osteoblasts

★CT = calcitonin

RENAL EXCRETION OF CALCIUM AND PHOSPHATE

This is summarised in Hawker, *Notebook of Medical Physiology: Renal and body fluids*, Chapter 2.

DISORDERS OF PARATHYROID FUNCTION

Hypoparathyroidism
This is due to insufficient secretion of PTH.

Hypo parathyroidism →
- ↓ plasma Ca^{++}
  - ↑ neuromuscular excitability → overt or latent tetany (positive Chvostek and Trousseau sign)
  - paraesthesiae (sensations of tingling, heat, etc.), usually precede tetany
  - laryngismus stridulus and convulsions (rarely)
- ↑ plasma phosphate

Special points
- tetany usually occurs 2–3 days after parathyroidectomy, and may occur as a transient or permanent complication of thyroidectomy if the parathyroid glands are damaged or removed
- spasm of laryngeal muscles → asphyxia if untreated
- in pseudohypoparathyroidism there is a genetic defect in cyclic AMP-mediated response to PTH at the cell level
- neonatal hypoparathyroidism may be associated with maternal hyperparathyroidism where increased maternal blood Ca suppresses neonate's parathyroids in utero

Note
- Ca^{2+} normally reduces membrane permeability of nerve and muscle fibres to Na^+
- ↓ Ca^{2+} (ECF) → ↑ membrane permeability to Na^+ → ↑ Na^+ influx → ↓ resting membrane potential → ↑ excitability → spontaneous discharge of nerve/muscle fibres → muscle spasm (tetany)
- tendency to tetany $\propto \dfrac{[HCO_3^-][HPO_4^{2-}]}{[Ca^{2+}][Mg^{2+}][H^+]}$

↓ $[H^+]$ → ↑ protein anion available to bind with Ca^{2+} → ↓ $[Ca^{2+}]$

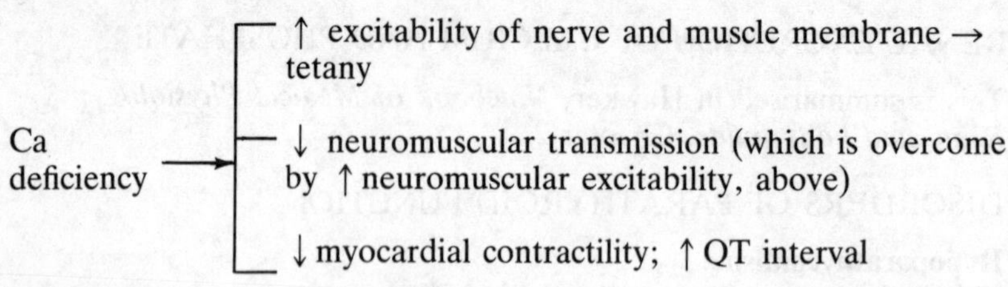

HYPERPARATHYROIDISM

This is due to excessive secretion of PTH.

1. Primary hyperparathyroidism
This is usually due to a parathyroid adenoma (85%), chief cell hyperplasia of 4 glands (14%), rarely carcinoma (1%).

Primary hyperparathyroidism →

- ↑ plasma Ca^{++} (e.g. 11 to 12 mg/100 ml, 2.75 to 3 mmol/l)
- ↓ plasma phosphate
- ↔ plasma alkaline phosphatase (↑ level if osteitis fibrosa cystica or associated bone disease present)
- ↑ osteoclastic activity → bone reabsorption → ↓ calcification of bone → osteitis fibrosa cystica (von Recklinghausen's disease)
- metastatic calcification
- hypercalciuria and renal stones
- effects of hypercalcaemia e.g. ↓ excitability of intestinal muscle membranes → ↓ muscle contractility → constipation; ↓ excitability of muscle membranes → ↓ muscle contractility → muscle weakness ; ↓ excitability of nerve membranes → ↓ reflex activity and depression of nervous activity; ↓ QT interval; ↓ relaxation of myocardium during diastole; thirst; renal damage (nephrocalcinosis) (nephrotoxic-effect → tubular medullary damage → ↓ H_2O reabsorption → polyuria); peptic ulcer; acute pancreatitis

The effects of excess plasma Ca are marked when the level reaches 16 mg/100 ml (~4 mmol/l).

2. Secondary hyperparathyroidism
This arises in response to hypocalcaemia.

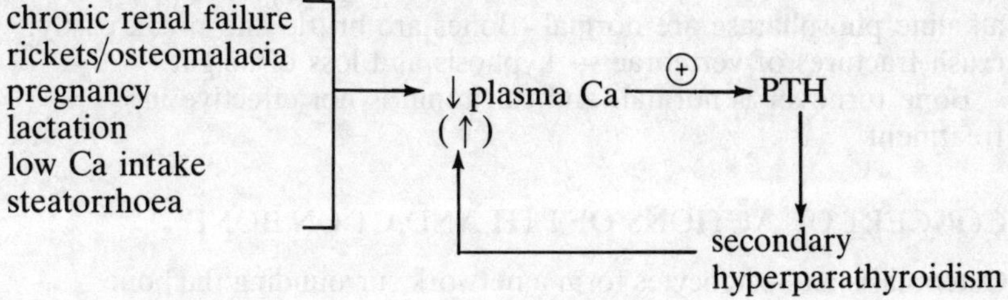

Prolonged stimulation of PTH by hypocalcaemia may rarely lead to autonomous excessive secretion of PTH: this is called tertiary hyperparathyroidism.

Rickets (or osteomalacia)
Calcium or phosphate deficiency in the plasma leads to reduced mineralisation of osteoid tissue and the bones become softened and weakened; but the organic matrix is normal.

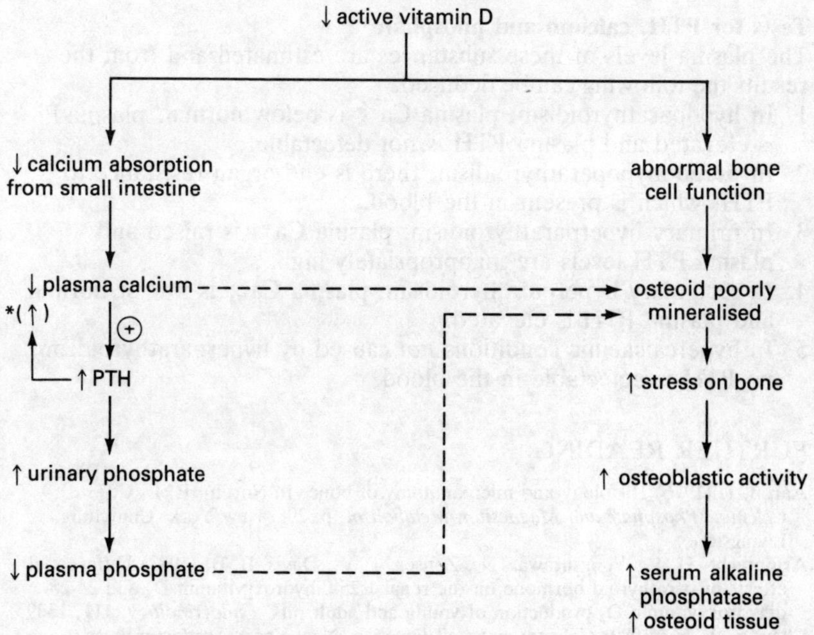

Fig. 10.5 Pathogenesis of rickets

OSTEOPOROSIS

The formation of organic matrix of bone is reduced; but the mineralisation of bone is normal. Plasma Ca, phosphate, and alkaline phosphatase are normal. Bones are brittle and break easily; crush fractures of vertebrae → kyphosis and loss of height.

Bone turnover is normal, and calcitonin is not effective in treatment.

CONCEPT OF ACTIONS OF PTH AND CT ON BONE

Osteoblasts and osteocytes form a network surrounding the bone matrix and separate the bone ECF from the general ECF. A calcium pump is located on the outer cell membrane and is thought to be stimulated by 1,25 dihydroxycholecalciferol and PTH. PTH increases, and CT decreases, the permeability of the inner cell membrane to calcium ions. 1,25 dihydroxychole-calciferol increases the pumping of calcium ions out of the cells into the general ECF and, thus, establishes a diffusion gradient for calcium movement from bone matrix → bone ECF → osteocyte/osteoblast → general ECF.

PARATHYROID FUNCTION TESTS (PRINCIPLES ONLY)

Tests for PTH, calcium and phosphate

The plasma levels of these substances are estimated and from the results the following can be deduced:

1. In hypoparathyroidism, plasma Ca^{++} is below normal, plasma Pi is elevated and plasma PTH is not detectable.
2. In pseudohypoparathyroidism, there is end-organ resistance to PTH, which is present in the blood.
3. In primary hyperparathyroidism, plasma Ca^{++} is raised and plasma PTH levels are inappropriately high.
4. In secondary hyperparathyroidism, plasma Ca^{++} is low or normal and plasma PTH is elevated.
5. In hypercalcaemic conditions *not* caused by hyperparathyroidism, no PTH is detectable in the blood.

FURTHER READING

Aaron, J. (1976) Histology and microanatomy of bone. In Nordin, B. E. C. (ed.) *Calcium, Phosphate and Magnesium Metabolism*, p. 298. New York: Churchill Livingstone.

Armbrecht, H. J., Wongsurawat, N., Zenser, T. V., Davis, B. B. (1982) Differential effects of parathyroid hormone on the renal 1,25-dihydroxyvitamin D_3 and 24-25-dihydroxyvitamin D_3 production of young and adult rats. *Endocrinology*, **111**, 1339.

Cheung, W. Y. (1980) *Calcium and Cell Function*. New York: Academic Press.

Copp, D. H., Talmage, R. V. (1978) *Endocrinology of Calcium Metabolism*. Proceedings of the Sixth Parathyroid Conference, Vancouver, Canada, June 12–17, 1977. Amsterdam: Excerpta Medica.

Deftos, L. J. (1978) Calcitonin in clinical medicine. In Stollerman, G. H. (ed.) *Advances in Internal Medicine*, vol. 23, p. 159–193. Chicago: Year Book Medical Publishers.

DeLuca, H. F. (1979) *Vitamin D. Metabolism and Function*. Berlin: Springer Verlag.

Dempster, D. W., Tobler, P. H., Olles, P., Born, W., Fischer, J. A. (1982) Potassium stimulates parathyroid hormone release from perifused parathyroid cells. *Endocrinology*, **111**, 196.

Greep, R. O., Astwood, E. B. (eds) (1976) *Handbook of Physiology*, vol. VII, Aurbach, G. D. (ed.) Parathyroid Gland, section 7, Endocrinology. Washington: American Physiological Society.

Habener, J. F., Kemper, B. W., Rich, A., Potts, J. T. Jr. (1977) Biosynthesis of parathyroid hormone. *Recent Progress in Hormone Research*, **33**, 249.

Hancox, N. M. (1972) *Biology of Bone: Biological Structure and Function*, vol. I. London: Cambridge University Press.

Lawson, D. E. M. (1978) *Vitamin D*. New York: Academic Press.

Nordin, B. E. C. (1976) *Calcium, Phosphate and Magnesium Metabolism*. New York: Churchill Livingstone.

Norman, A. W. (1979) *Vitamin D*. London: Academic Press.

Parsons, J. A. (1979) Physiology of parathyroid hormone. In DeGroot, L. J. et al (eds) *Endocrinology*, vol. 2, p. 621. New York: Grune & Stratton.

Scarpa, A., Carafoli, E. (1978) *Calcium Transport and Cell Function*, p. 307. New York: Academic Press.

Shinki, T., Takahashi, N., Kawate, N., Suda, T. (1982) The possible role of calcium-binding protein induced by 1α, 25-dihydroxyvitamin D_3 in the intestinal calcium transport mechanism. *Endocrinology*, **111**, 1546.

Multiple choice questions

1. In patients with hypercalcaemia:
1. administration of sodium phosphate intravenously will cause a further rise in serum calcium;
2. the level of serum phosphate gives a clear indication of the likely cause;
3. urinary calcium excretion is relatively lower than expected from the serum calcium level in patients with primary hyperparathyroidism;
4. the renal effects of PTH are significant in leading to the elevated calcium level of primary hyperparathyroidism;
5. administration of thiazide diuretics may worsen the hypercalcaemia;
6. there may be a demonstrable impairment in renal concentrating ability.

2. With respect to bone and calcium metabolism:
1. chronic calcitonin excess in man does not usually cause hypocalcaemia;
2. plasma calcitonin is markedly elevated in primary hyperparathyroidism;
3. thickened osteoid seams are indicative of osteomalacia;
4. increased osteoclastic activity elevates serum alkaline phosphatase;
5. hypocalcaemia does not stimulate osteoclasts directly;
6. vitamin D in high doses elevates plasma Ca^{++} even in the total absence of the parathyroid glands;
7. plasma PTH is raised in immobilisation hypercalcaemia;
8. urinary hydroxyproline is an index of bone resorption;
9. urinary hydroxyproline is raised during the growing period because of increased bone turnover;
10. the diameter of long bones decreases with age.

3. Hyperparathyroidism is generally associated with:
1. osteomalacia;
2. osteitis fibrosa cystica;
3. radio-opaque renal calculi;
4. a benign tumour of the parathyroid glands;
5. cramps and tetany;
6. a defect in water concentrating ability.

4. With respect to PTH secretion:
1. it responds directly to changing levels of plasma Ca^{++};
2. it responds indirectly to changing levels of plasma Mg^{++};
3. it responds directly to changing levels of plasma P_i;
4. it is increased in chronic renal failure;
5. increased secretion is not necessarily accompanied by elevated plasma Ca^{++};

6. it is inhibited by chronic Mg^{++} deficiency;
7. the elevated plasma P_i of hypoparathyroidism is caused by increased release of P_i from bone;
8. it is inhibited by the hypercalcaemia of vitamin D intoxication.

5. Hypercalcaemia may be a complication of:
1. carcinoma of the lung without metastases in bone;
2. carcinoma of the kidney without metastases in bone;
3. sarcoidosis;
4. parathyroidectomy;
5. vitamin D therapy;
6. immobilisation in bed;
7. pancreatitis.

6. In hyperparathyroidism:
1. serum calcium is invariably elevated;
2. myopathy can be severe;
3. pseudofractures may occur;
4. parathyroid adenomata may be found in the posterior mediastinum;
5. if multiple adenomata are present, other endocrine disease should be suspected.

7. With regard to parathyroid function:
1. serum parathormone (PTH) levels are characteristically elevated in both primary and secondary hyperparathyroidism;
2. primary hyperparathyroidism is associated with a tendency to hyperchloraemic acidosis because of the effects of PTH on the kidney;
3. the biologically active portion of PTH is located at the C-terminal end of the molecule;
4. PTH increases the conversion of 25-OH cholecalciferol to the active 1,25 di-OH cholecalciferol in the kidney;
5. administration of 40 mg prednisolone daily for 10 days usually causes no change in serum calcium levels in patients with primary hyperparathyroidism.

8. In patients with renal calculi:
1. in more than 50% of cases, the aetiology is unknown;
2. cystine stones are radio-lucent;
3. primary hyperoxaluria is the usual cause of calculi in patients whose stones contain calcium oxalate;
4. allopurinol is a useful agent in the prevention of renal stone formation in patients with gout;
5. hydrochlorothiazide is useful in the treatment of idiopathic hypercalciuria.

Answers

1. 3, 4, 5, 6
2. 1, 3, 5, 6, 8, 9
3. 2, 3, 4, 6
4. 1, 4, 5, 6, 8
5. 1, 2, 3, 5, 6
6. 2, 4, 5
7. 1, 2, 4, 5
8. 1, 4, 5

11. Maternal physiology

CARDIOPULMONARY SYSTEM

1. Cardiovascular system

As metabolic, thermal, acid-base and other stresses are placed on the maternal organism during pregnancy there is a need for an increased cardiac output, blood volume and blood flow to many tissues and organs.

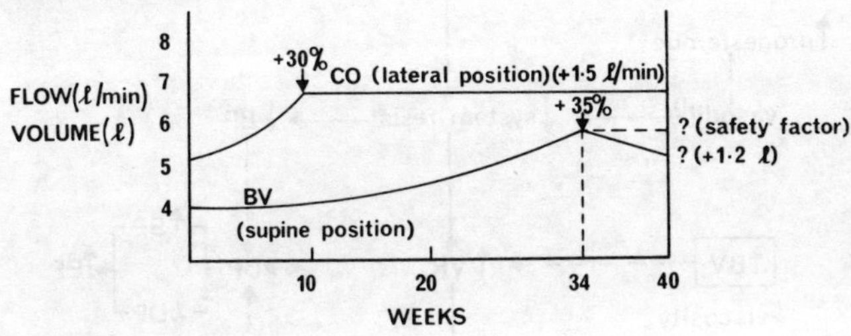

Fig. 11.1 Cardiac output and blood volume during pregnancy
Key: CO = cardiac output
BV = blood volume

a. Blood flow

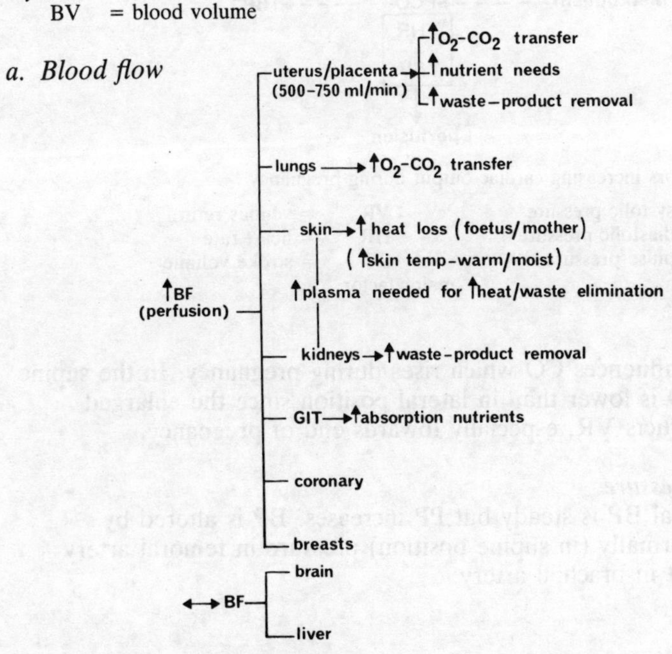

Fig. 11.2 Distribution of increased blood flow during pregnancy

b. Cardiac output

↑ CO is required for ↑ perfusion, and is met by ↑ HR, ↑ SV (CO = HR.SV

↑ CO maintains mean BP in the presence of a large A-V shunt (placenta).

↑ BV maintains vascular filling in presence of increased vascular bed volume (placenta, skin).

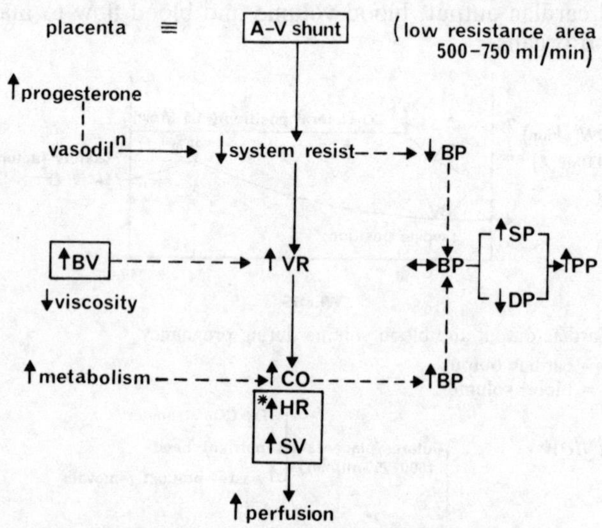

Fig. 11.3 Factors increasing cardiac output during pregnancy

Key: SP = systolic pressure VR = venous return
DP = diastolic pressure HR = heart rate
PP = pulse pressure SV = stroke volume
* major factor

Posture influences CO which rises during pregnancy. In the supine position CO is lower than in lateral position since the enlarged uterus obstructs VR, especially towards end of pregnancy.

c. Blood pressure

Mean arterial BP is steady but PP increases. BP is altered by posture. Normally (in supine position) pressure in femoral artery exceeds that in brachial artery.

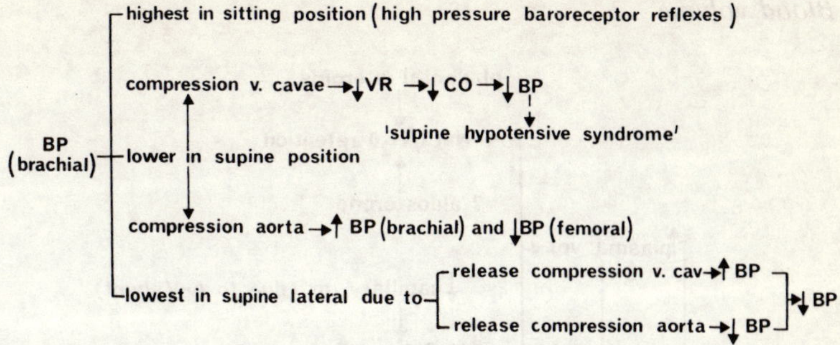

Fig. 11.4 Effects of posture on blood pressure during pregnancy

Key: $PR = \frac{BP}{CO}$

In supine lateral position, v. cava is still partially compressed and this tends to ↓ BP.

PR decreases during pregnancy, but the BP is steady because the CO increases.

d. Venous pressure

Femoral venous pressure progressively increases as pregnancy advances, reaching approximately 20 mmHg at term (supine position). Central venous pressure does not change.

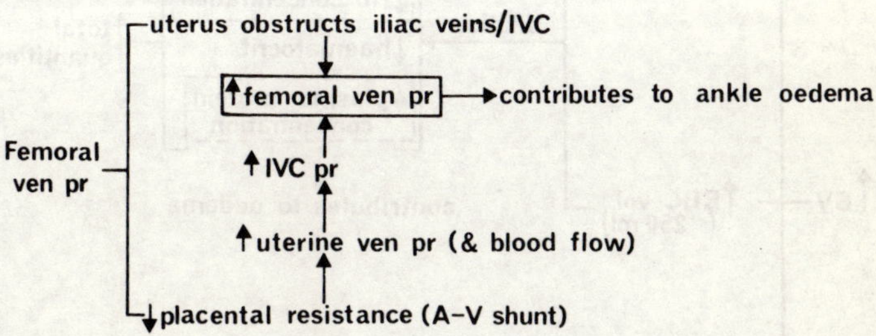

Fig. 11.5 Factors influencing femoral venous pressure during pregnancy

e. Blood volume

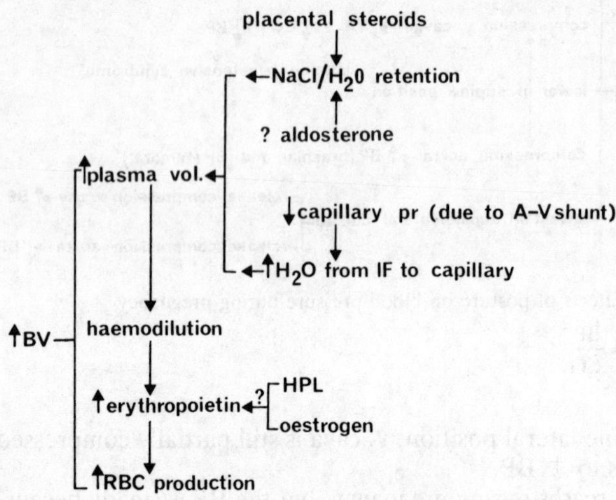

Fig. 11.6 Mechanism of increased blood volume during pregnancy

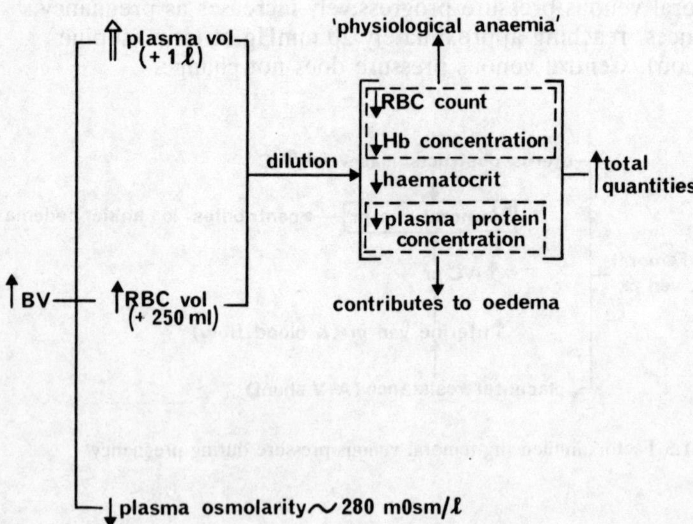

Fig. 11.7 Effects of increased blood volume during pregnancy

During pregnancy the O_2 consumption/min increases. In early pregnancy the CO increases proportionately more than the O_2 consumption and hence the A-V O_2 difference is small. Only in late pregnancy does this difference equal (or just exceed) that in nonpregnancy; so that in pregnancy more O_2/100 ml of venous blood returns to the heart unused.

2. Respiratory system

Minute ventilatory volume increases during pregnancy and there is an increased O_2 consumption/min.

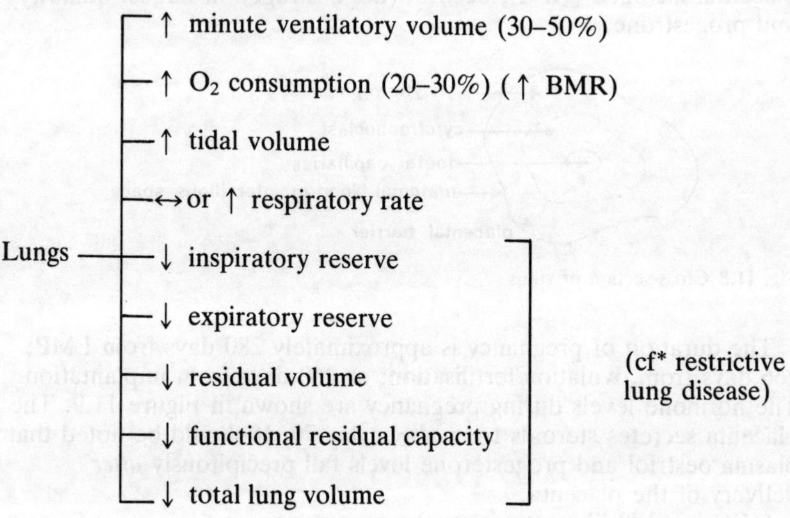

↓ movement of diaphragm (hindered by ↑ abdominal contents, especially in supine position) restricts expansion of lung.

ENDOCRINOLOGY OF PREGNANCY

Fertilisation usually occurs in the midportion of a Fallopian tube. Sperm penetration of the ovum is facilitated by lysosomal enzymes (hyaluronidase and proteinase) secreted by acrosomes. Normally only one sperm penetrates the ovum: a protective barrier forms which prevents other sperm from entering. The tail breaks off and the nuclei fuse forming a genotype of 46 chromosomes. Cell division results in the blastocyst which is propelled towards the uterus by ciliary movements and peristalsis of the Fallopian tubes, and is

* = see Hawker, *Notebook of Medical Physiology: Cardiopulmonary*, Chapter 9

nourished by secretions induced by progesterone. An outer layer of cells — the trophoblast — forms during this time. About 7 days after fertilisation it implants in the succulent endometrium apparently due to erosion by enzymes secreted by syncytiotrophoblast cells. The cytotrophoblast and/or syncytiotrophoblast cells almost immediately begin to secrete human chorionic gonadotrophic hormone (HCG).

The site of production of the placental hormones is not established, however ultrastructure, histochemical and immunofluorescent studies implicate the syncytiotrophoblast cells in their production. As well as HCG, these hormones include human placental lactogen (HPL), oestriol (the oestrogen in largest quantity) and progestrone.

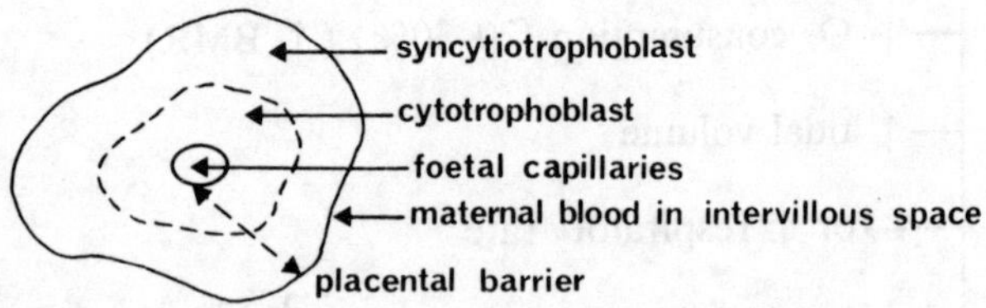

Fig. 11.8 Cross-section of villus

The duration of pregnancy is approximately 280 days from LMP; 266 days from ovulation/fertilisation; or 260 days from implantation. The hormone levels during pregnancy are shown in Figure 11.9. The placenta secretes steroids from about day 70. It should be noted that plasma oestriol and progesterone levels fall precipitously *after* delivery of the placenta.

HCG has LH-like actions on the corpus luteum.

HPL structurally resembles human growth hormone and has prolactin and growth hormone activities. It is a protein (190 amino acids).

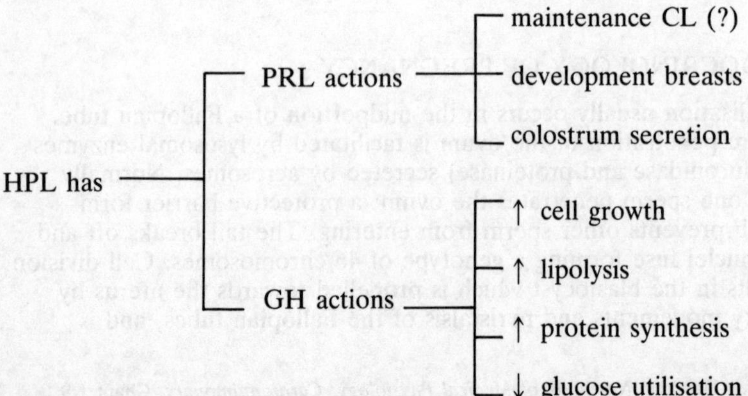

Hormones in pregnancy

The developing blastocyst secretes HCG which, during the first trimester, stimulates growth of the corpus luteum and secretion of oestrogen and progesterone. These ovarian steroids maintain: the developing/differentiating fetoplacental unit in situ; the inhibition of FSH and LH secretion ('contraceptive effect'); and the growth of uterus, breasts and other structures.

During the first trimester removal of the corpus luteum produces abortion. After 3 months both ovaries can be removed without interruption of pregnancy, since the fetoplacental unit functions autonomously and secretes increasing quantities of steroids (previously produced by the corpus luteum). These steroids maintain suppression of secretion of FSH and LH and maintain the functional integrity of the fetoplacental unit. It is postulated that HPL acting via the hypothalamus suppresses secretion of prolactin (and possibly growth hormone).

During pregnancy, ovulation and lactation are held in abeyance (colostrum may be secreted at 5th month).

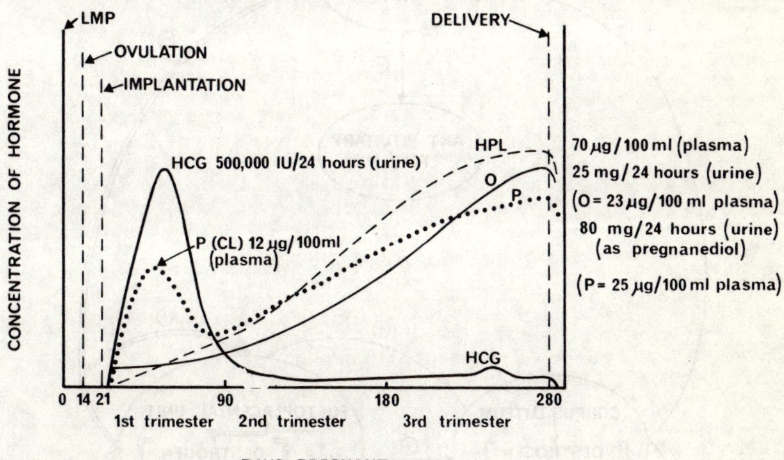

Fig. 11.9 Hormone levels during pregnancy

Key: HCG = human chorionic gonadotrophin
HPL = human placental lactogen
O = oestriol
P = progesterone
CL = corpus luteum
LMP = last menstrual period

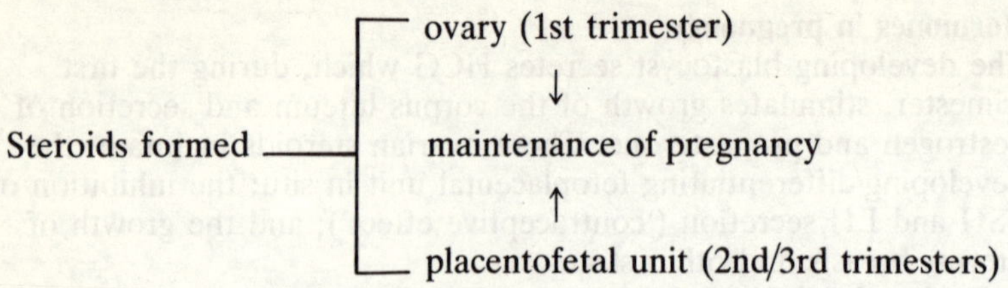

The corpus luteum begins to degenerate about the 5th month, but it persists throughout pregnancy.

HCG levels are used in pregnancy tests. The rising HCG production in early pregnancy allows a positive result in the immunological methods used from about 26 days after conception.

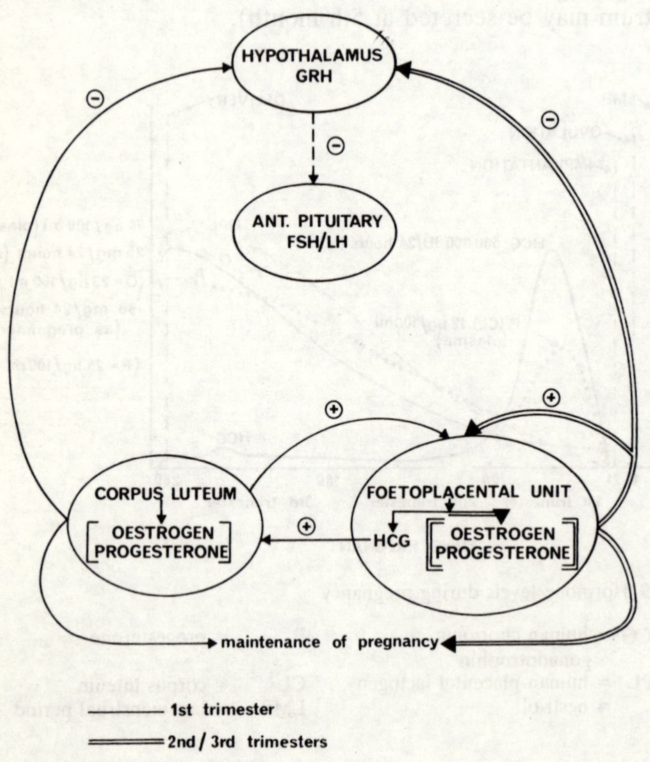

Fig. 11.10 Hormones in maintenance of pregnancy

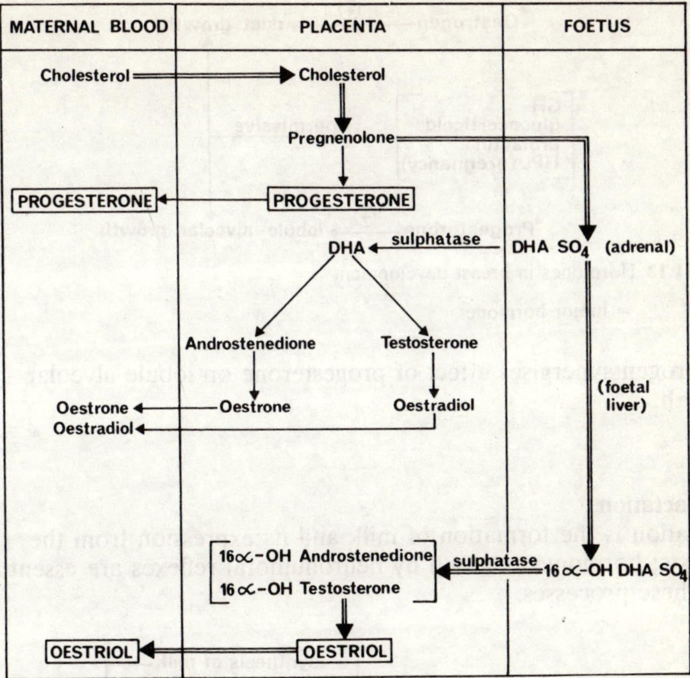

Fig. 11.11 Fetoplacental unit and steroid biosynthesis

Key: DHA = Dehydroepiandrosterone (sulphate)

Balanced enzyme system between placenta and fetus for synthesis of oestriol which quantitatively is the most important oestrogen. The placenta possesses aromatising enzymes but lacks C16 hydroxylase. Plasma oestriol levels correlate with fetal/placental function and hence are predictive of state of fetal health.

BREAST DEVELOPMENT AND LACTATION

1. Mammogenesis

Breast development at puberty is hormone-dependent; oestrogens and progesterone are essential, and other hormones influence differentiation, development, and secretory activity of cells.

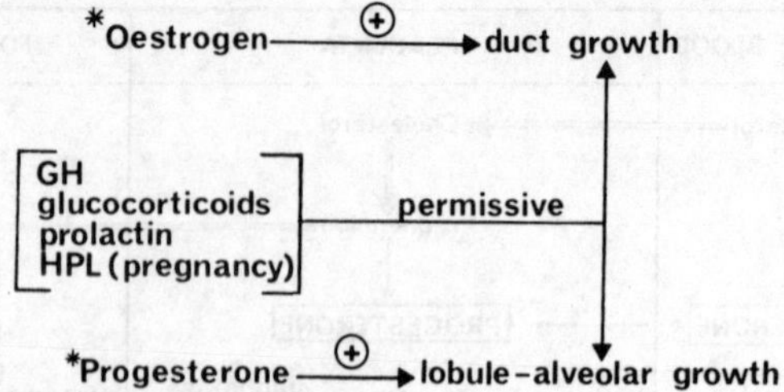

Fig. 11.12 Hormones in breast development

Key: * = major hormones

Oestrogen synergises effect of progesterone on lobule-alveolar growth.

2. Lactation

Lactation is the formation of milk and its expression from the breasts; hormones released by neurohumoral reflexes are essèntial for these processes.

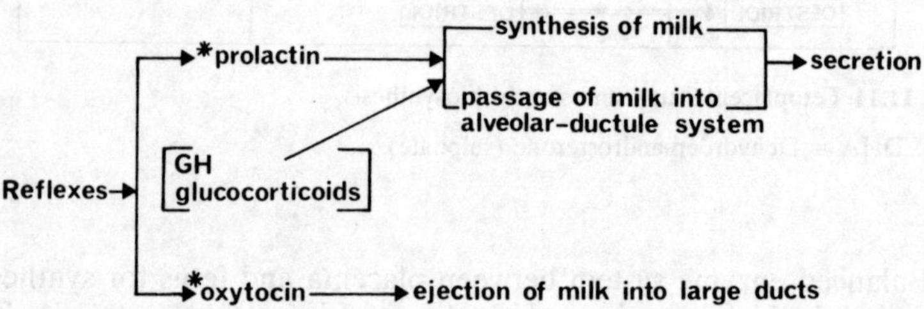

Fig. 11.13 Secretion and ejection of milk

Key: * = major hormones

'Milk' (colostrum) is secreted normally after about 16 weeks of pregnancy and distension of the breasts occurs. Abortion after this time is followed by some 'milk' production.

There are two aspects of lactation:

a. Initiation of lactation
b. Maintenance of lactation.

Milk production commences 1 to 3 days after parturition. Initially colostrum is secreted which has a low fat content.

Suckling maintains the formation and release of milk. Oestrogen administration suppresses lactation.

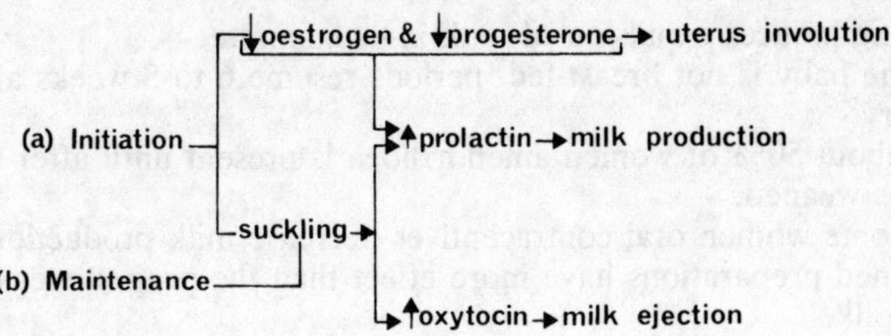

Fig. 11.14 Lactation

The composition of milk becomes modified during the initial few weeks of nursing. When lactation is firmly established about 1.5 l of milk are produced daily. Casein and lactalbumin are the main proteins and little iron is present.

Table 11.1 Composition per litre

	Milk	Colostrum
Calories	750	650
Calcium (g)	0.35	0.5
Iron (mg)	0.5	1.0
Protein (g)	10	20
Lactose (g)	70	60
Fat (g)	45	30

Nursing the baby induces a number of integrated responses.

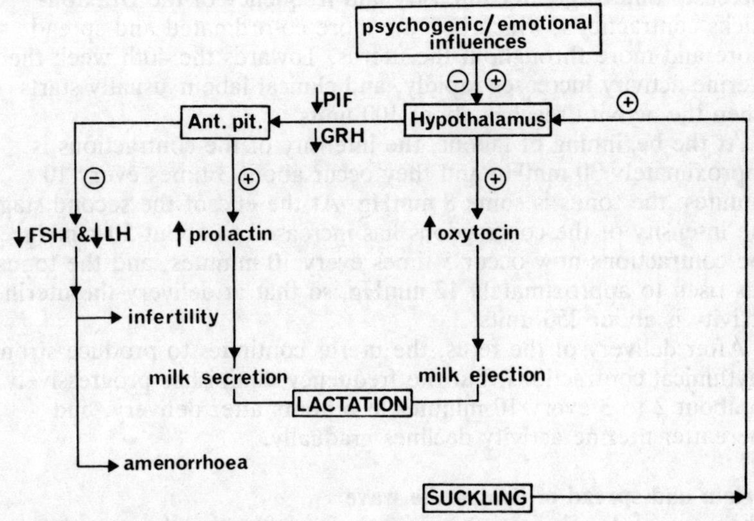

Fig. 11.15 Responses to nursing

Suckling induces expulsion of milk in ~1 minute.

If the baby is not breast-fed, periods resume 6 to 8 weeks after delivery.

In about 50% of women amenorrhoea is present until after the baby is weaned.

In some women oral contraceptives decrease milk production. The combined preparations have more effect than the progestogen-only 'mini pill'.

UTERINE CONTRACTILITY DURING PREGNANCY AND LABOUR

Uterine activity can be measured in arbitrary units by recording the intra-amniotic pressure or intrauterine pressure (i.e. the intensity) and frequency of the contractions. At the onset of labour, if the contractions have an intensity of 40 mmHg and a frequency of 3 per 10 minutes, then the uterine activity is 3 × 40 (120) Montevideo units.

Throughout pregnancy and labour there is an evolution of spontaneous uterine contractility. During the first 30 weeks the uterus is relatively quiet. The contractions are small, of several mmHg intensity, occur about once per minute and are localised to small areas of the uterus. Every half-hour or so Braxton-Hicks contractions of about 10 to 15 mmHg intensity interrupt these small contractions and spread over a larger area of the uterus.

After the 30th week of gestation, the uterine activity progressively increases due to greater intensity and frequency of the Braxton-Hicks contractions, which become more co-ordinated and spread more and more throughout the uterus. Towards the 40th week the uterine activity increases rapidly, and clinical labour usually starts when the activity reaches about 100 units.

At the beginning of labour, the intensity of the contractions is approximately 30 mmHg and they occur about 3 times every 10 minutes: the tonus is some 8 mmHg. At the end of the second stage, the intensity of the contractions has increased to about 50 mmHg, the contractions now occur 5 times every 10 minutes, and the tonus has risen to approximately 12 mmHg, so that at delivery the uterine activity is about 250 units.

After delivery of the fetus, the uterus continues to produce strong rhythmical contractions, but the frequency diminishes progressively to about 2 to 3 every 10 minutes at 6 hours after delivery, and thereafter uterine activity declines gradually.

Origin and spread of contractile wave

The wave of depolarisation originates from functional pacemaker cells near the uterine end of one Fallopian tube and spreads across the fundus and downwards to the cervix. The wave is so well coordinated that all parts of the uterus contract maximally at the same time, and the upper part of the uterus contracts more strongly than the lower part and cervix.

Functions of uterine contractions during labour

1. Dilatation of the cervix — the more powerful contractions of the upper segment exert a longitudinal traction on the cervix causing its progressive effacement and dilatation. Retraction occurs as the upper segment becomes thicker and shorter after each contraction.
2. Descent and delivery of the fetus.
3. Separation of the placenta.
4. Placentofetal transfusion 70 to 90 ml blood.
5. Haemostasis.

Oxytocin during pregnancy and labour

The level of oxytocin increases up to midpregnancy and then plateaus at this high level throughout the remainder of pregnancy and during dilatation of the cervix; the level rises during the second stage of labour. The oxytocinase level in the plasma mirrors reasonably closely the oxytocin level up to the onset of labour.

The sensitivity of the uterus progressively increases up to 35th week of pregnancy, and from then on and during labour its sensitivity to oxytocin remains unchanged.

Mechanism of labour

The factors which initiate and maintain natural labour are unknown. Various uterine stimulants and depressants have been identified and, against this background knowledge, a number of concepts on the mechanism of labour have been framed.

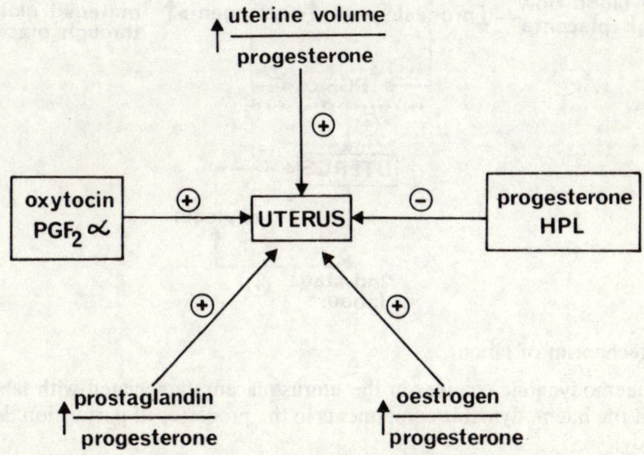

Fig. 11.16 Factors influencing activity of uterus

Oxytocin and $PGF_2\alpha$ are potent uterine stimulants, and progesterone is a depressant of uterine contractility. HPL also depresses uterine contractility, especially in the presence of progesterone.

The activity of the uterus is influenced by the uterine volume/progesterone plasma concentration ratio (activity increases as ratio increases), and by the prostaglandin/progesterone plasma concentration ratio (activity increases as ratio increases).

The sensitivity of the uterus to oxytocin and to $PGF_2\alpha$ is modulated by the oestrogen/progesterone concentration ratio (sensitivity increases as ratio increases).

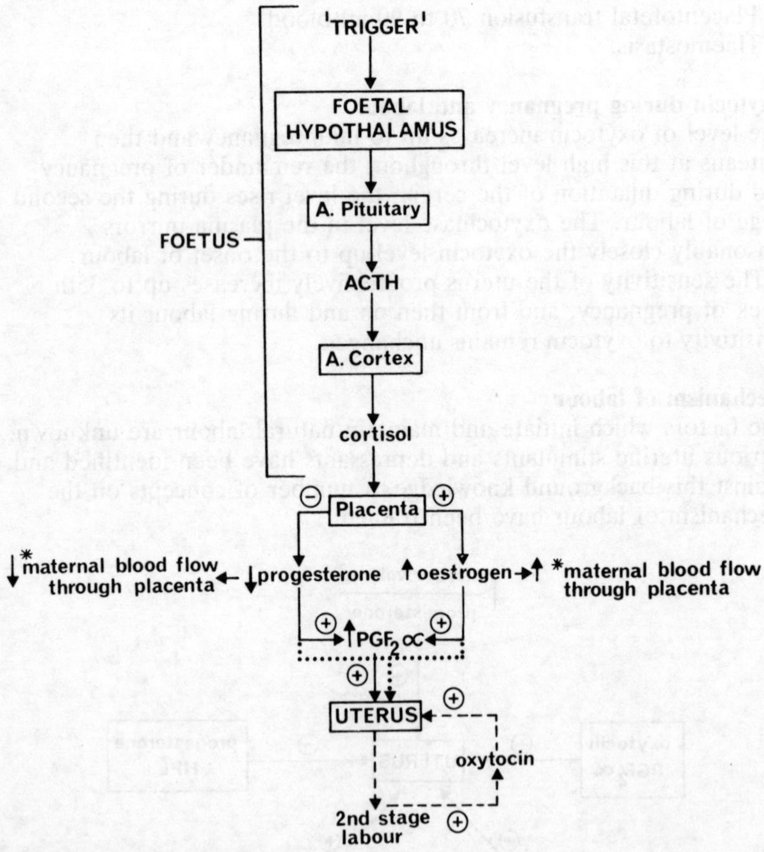

Fig. 11.17 Mechanism of labour

Key: * The haemodynamic changes in the uterus/placenta associated with labour are complex, and the haemodynamic components in the processes of parturition deserve intensive study.

. . . . : ↑ $\frac{\text{oestrogen}}{\text{progesterone}}$ maintains high sensitivity of uterus to $PGF_2\alpha$ and oxytocin

Oestradiol and increased uterine volume enhance the prostaglandin content in the uterus and placenta; and progesterone decreases prostaglandin production.

Recent studies in women suggest that plasma levels of progesterone and HPL decrease over the last month of pregnancy, and that oestradiol and $PGF_2\alpha$ increase concurrently. At this time the level of oxytocin in the blood is high and the uterus is now exquisitely sensitive to oxytocin.

Hence the hormonal environment of the uterus is conducive to its increased activity at term.

Combining observations made in animal studies with those in the human, Figure 11.17 presents one concept of the mechanism of labour in which a biological 'clock' in the fetus initiates the hormonal events.

FURTHER READING

Anderson, R. R. (1974) Endocrinological control. In Larson, B. L., Smith, V. R. (eds) *Lactation. A Comprehensive Treatise*, p. 97. New York: Academic Press.

Friesen, H. G., Tolis, G. (1977) The use of bromocriptine in the galactorrhea-amenorrhea syndromes: the Canadian cooperative study. *Clinical Endocrinology* (Oxf.), **6** (Suppl.), 91s.

Goldberg, V. J., Ramwell, P. W. (1975) Role of prostaglandins in reproduction. *Physiological Reviews*, **55**, 325–351.

Klopper, A., Gardner, J. (eds) (1973) Endocrine factors in labour. *Memoirs. Society for Endocrinology*, **20**, 1–162.

Liggins, G. C. (1979) Initiation of parturition. *British Medical Bulletin*, **35**, 145.

Nathanielsz, P. W. (1978) Endocrine mechanisms of parturition. *Annual Review of Physiology*, **40**, 411.

Pearson, J. F., Weaver, J. B. (1976) Fetal activity and fetal well-being: An evaluation. *British Medical Journal*, **1**, 1305.

Rose, J. C., Meis, P. J., Urban, R. B., Greiss Jr, F. C. (1982) In vivo evidence for increased adrenal sensitivity to adrenocorticotropin-(1–24) in the lamb fetus late in gestation. *Endocrinology*, **111**, 80.

Vorherr, H. (1974) Suppression of lactation. In Vorrherr, H. (ed.) *The Breast: Morphology, Physiology, and Lactation*, p. 198. New York: Academic Press.

Multiple choice questions

1. During the *last* 3 months of pregnancy in women:
   1. oestriol is produced by the placenta following synthesis of precursor substances by the fetal adrenal cortex;
   2. oestriol is produced by the placenta and the fetus is not involved in this synthesis;
   3. oestradiol is the oestrogen in greatest quantity in the maternal blood;
   4. progesterone is produced by the placenta and the fetus is not involved in this synthesis;
   5. progesterone is produced mainly by the corpus luteum;
   6. chorionic gonadotrophic hormone (HCG) reaches its highest level in the maternal blood;
   7. human placental lactogen is at a higher level in the maternal blood than during the first 3 months;
   8. removal of both ovaries would be expected to induce abortion.

2. During the *first* 3 months of pregnancy in women:
   1. removal of both ovaries is followed by abortion;
   2. growth of the corpus luteum is due mainly to HCG;
   3. oestriol is at a higher level in the maternal blood than during the last 3 months;
   4. the corpus luteum rapidly regresses;
   5. the secretion into the maternal blood of FSH and LH increases.

3. High serum thyroxine levels in pregnancy are due to:
   1. increased hormone synthesis;
   2. reduced renal clearance;
   3. fetal thyroxine secretion;
   4. increased thyroxine binding globulin;
   5. increased maternal blood volume.

4. The hormones required for breast development (i.e. mammogenesis) in women are:
   1. oestrogen for duct growth;
   2. progesterone alone for duct growth;
   3. progesterone plus oestrogen for lobule-alveolar growth;
   4. oestrogen alone for lobule-alveolar growth;
   5. prolactin alone for both duct growth and lobule-alveolar growth;
   6. growth hormone plus cortisol acting together (and without oestrogen) for duct growth.

5. With respect to oestriol assay as a method of assessment of fetoplacental function:
1. in human pregnancy, oestriol is produced almost exclusively by placental aromatisation of fetal precursor substances;
2. deficiency of the fetal pituitary, as in anencephaly, results in bilateral hypertrophy of the fetal adrenal cortex and elevated levels of oestriol precursor substances;
3. corticosteroids given to the mother suppress fetal adrenocortical function and cause reduced oestriol production;
4. in obtaining a 24-hour urinary collection for oestriol measurement it is important to maintain a urinary excretion of at least 2 litres per day;
5. certain drugs given to the mother in pregnancy, including ampicillin, cause depression of maternal urinary oestriol excretion.

6. Concerning lactation in women:
1. milk secretion is initiated by increased prolactin secretion;
2. milk ejection is due to prolactin;
3. lactation is often associated with infertility which results from decreased follicle-stimulating hormone (FSH) and luteinising hormone (LH) production;
4. lactation which is maintained by suckling, is due to increased LH production;
5. milk secretion, which follows abortion after the fifth month of pregnancy, is due to increased prolactin secretion.

7. Iron deficiency anaemia in pregnancy:
1. is more common in first than subsequent pregnancies;
2. is usually associated with a raised total iron-binding capacity in addition to a low level of serum iron;
3. has a characteristic microscopic appearance of microcytosis and hypochromia of the erythrocytes;
4. treated with daily intramuscular injections of iron dextran may be expected to show an increase in haemoglobin levels in about 7 days;
5. requires accurate differential diagnosis since thalassaemia is a contraindication to iron therapy.

8. In the lecithin/sphingomyelin ratio test for assessment of fetal lung maturity:
1. lecithin, measured in this test, is produced in the alveoli of the fetal lung;
2. sphingomyelin has the same origin as lecithin;

3. the lecithin/sphingomyelin ratio is measured in maternal urine;
4. the test is invalidated if the specimen is contaminated with blood;
5. the ratio of >2 is generally taken to indicate mature fetal lungs and little likelihood of respiratory distress syndrome in the newborn.

9. During pregnancy:
1. the intensity of uterine contractions in first-stage labour is about 30 to 50 mmHg;
2. the sensitivity of the uterus to oxytocin is greater at 38 weeks than at 20 weeks;
3. when uterine contractions are well co-ordinated in first-stage labour the contractile wave originates in the lower uterine segment;
4. the intensity of the uterine contractions during second-stage labour can approximate that of the peristalic contractions of the stomach;
5. the intrauterine pressure during second-stage labour can approach maternal systolic BP;
6. after 20 weeks the fetal kidneys produce a significant part of the amniotic fluid;
7. near term the volume of amniotic fluid is about 500 to 1000 ml;
8. near term a lecithin/spingomyelin ratio >2 in amniotic fluid suggests maturation of the lungs with respect to synthesis of surfactant;
9. the level of HCG in the maternal blood is higher at 30 weeks than at 7 weeks.

10. A week before birth in full-term pregnancy the maternal:
1. cardiac output is higher than at 4 weeks of pregnancy;
2. blood volume is higher than at 20 weeks of pregnancy;
3. blood flow/min through the skin is generally higher than at 6 weeks of pregnancy;
4. blood flow/min through the brain is higher than at 4 weeks of pregnancy;
5. blood flow/min through the coronary vessels is higher than at 4 weeks of pregnancy;
6. corpus luteum synthesises significant quantities of oestrogen;
7. supine position can lead to hypotension, due to compression of the inferior vena cava by the enlarged uterus;
8. supine lateral position can lead to a lower brachial artery pressure than in the supine position because, when in the former position, partial compression of the inferior vena

cava and descending aorta by the enlarged uterus is released;

9. blood volume is higher than at 10 weeks of pregnancy and the red cell volume is increased proportionately more than the plasma volume so that haemoconcentration occurs;
10. alveolar ventilation/min is higher than at 5 weeks of pregnancy because the maternal $PaCO_2$ is raised;
11. O_2 consumption/min is higher than at 5 weeks of pregnancy because of mild hyperthyroidism.

Answers

1. 1, 4, 7
2. 1, 2
3. 4
4. 1, 3
5. 1, 3, 5
6. 1, 3, 5
7. 2, 3, 5
8. 1, 4. 5
9. 1, 2, 4, 5, 6, 7, 8
10. 1, 2, 3, 5, 7, 8

---

12. Fetal physiology

SEX DETERMINATION, DIFFERENTIATION AND DEVELOPMENT

The chromosomal constitution is determined at conception, and the genotype determines the gonadal sex. The intrinsic pattern of genital differentiation is female; intervention of male duct organisers and androgens (both elaborated by the testis) is necessary for the male pattern of internal and external genital differentiation.

Until the 7th week the gonad is bipotential and consists of a primitive medulla and cortex. In the absence of the Y chromosome, sex determining genes carried on the X chromosomes cause an ovary to form from the indifferent gonad. Inductor substances produced by the differentiating cortical component stimulate differentiation of the cortex and inhibit development of the medulla.

Two X chromosomes are necessary for normal ovarian development. In the presence of only one X chromosome (Y absent) the indifferent gonad fails to develop (agenesis).

In the XY karyotype fetus, sex determining genes carried on the Y chromosome induce differentiation of the medulla, and this tissue produces inductors which stimulate the growth of the medulla and inhibit development of the cortex. The differentiating testis Sertoli cells) secretes duct organisers which act locally on the Wolffian structures and cause them to develop into male internal structures (epididymis, vas deferens, seminal vesicles and ejaculatory duct) and cause regression of the Müllerian structures which otherwise would develop into Fallopian tubes and uterus.

The testis secretes testosterone which induces male development of the external genitalia. Under the influence of androgen the primitive genital tubercle develops into a penis (rather than clitoris); the urethral folds, into corpus spongiosum (rather than labia minora); the labioscrotal swellings, into scrotum (rather than labia majora); the urogenital sinus, into bulbourethral glands and prostatic urtricle (rather than outer ⅔ vagina and Bartholin's and Skene's glands).

If, with the XY karyotype, the testis fails to secrete organisers which are vital for duct transformations, or if the primordial structures are unresponsive to organisers adequately produced, the internal structures develop according to the female pattern.

If a testis fails to develop on one side, female internal structures develop ipsilaterally: male internal genitalia develop on the side with the normal testis.

Concept
The indifferent gonad, and primordial internal ducts and external structures would become feminised if not diverted by masculinising mechanisms.

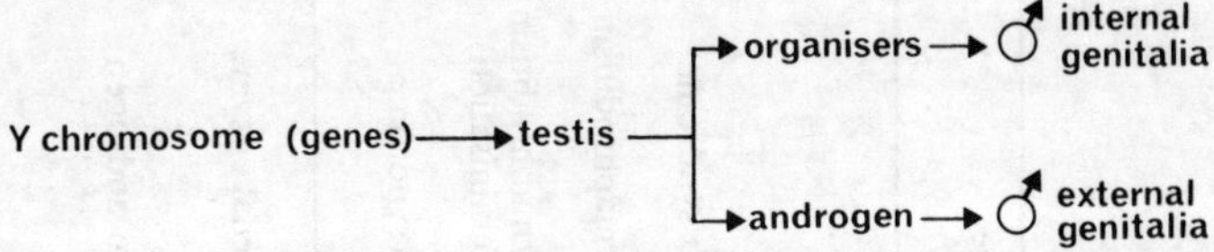

1. Sexual differentiation of reproductive system

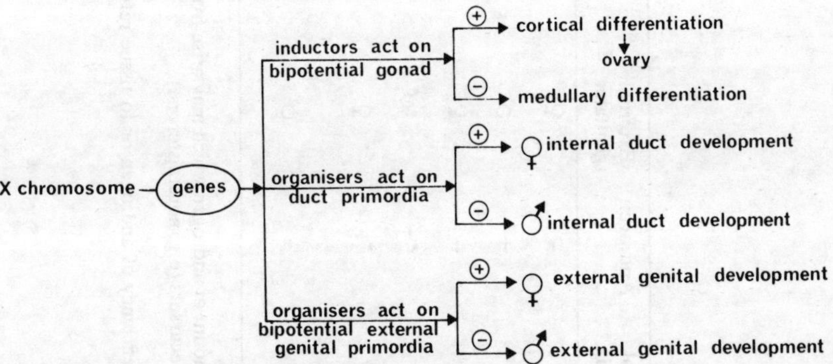

Fig. 12.1 Differentiation of female reproductive tract

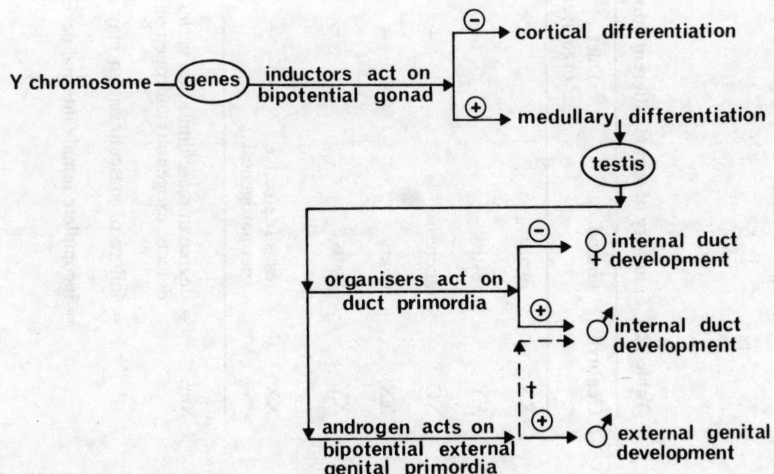

Fig. 12.2 Differentiation of male reproductive tracts

Key: † = androgen secreted by Leydig cells has minor supportive effect on ♂ duct development, and no effect on ♀ duct system

Table 12.1 Summary of sexual differentiation

Genotype	Gonad	Duct organiser	Internal genitalia	Androgen	External genitalia	Clinical stage
XX	ovary	–	♀	–	♀	♀
XY	testis	+	♂	+	♂	♂
XO	agenesis	–	♀	–	♀	Turner's syndrome
XX	ovary	–	♀	*+	*♂	♀Pseudohermaphroditism
XY	testis	–	‡♀	†+/–	♀	♂Pseudohermaphroditism (Testicular feminisation)
XXY	testis (defective spermatogenesis)	+	♂	+	♂	Klinefelter's syndrome

Key: * = degree of masculinisation varies with the amount of androgen which may arise from; fetal adrenal enzyme defects, exogenous androgen or androgen precursors (e.g. norethisterone)

† = failure of masculinisation may be due to deficiency of androgen or to tissue refractoriness to androgen

‡ = incomplete female internal development

2. Sexual differentiations of hypothalamus
Studies in animals including primates show that structures in the brain differentiate sexually in late embryonic or early neonatal life so that normally during reproductive life, the secretion of gonadotrophic hormones and the psychosexual behaviour of the animal are appropriate to its genetic sex.

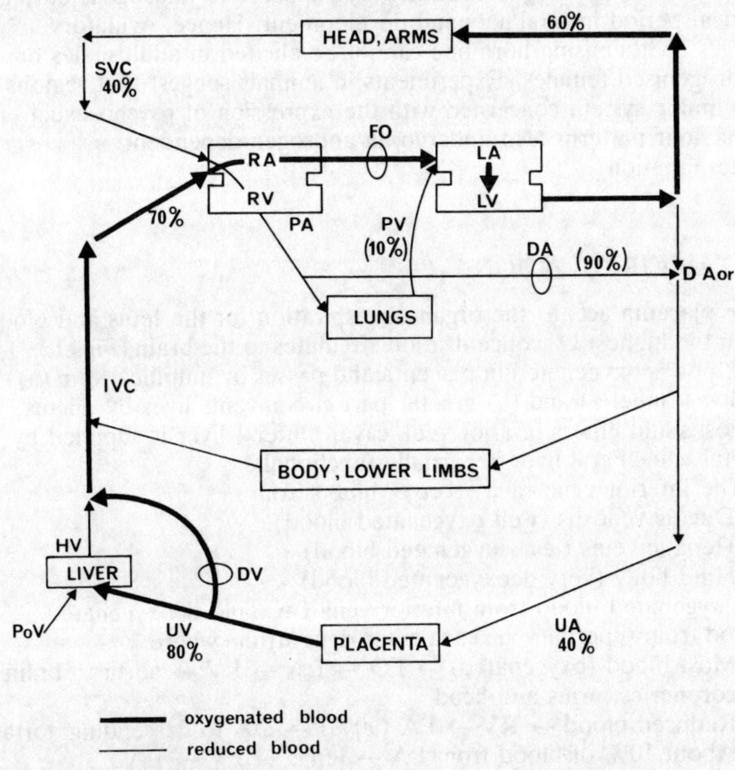

Fig. 12.3 Fetal circulation

% = O_2 saturation
(%) = flow through PV and DA

Key:

RA	= right atrium	HV	= hepatic vein
RV	= right ventricle	PoV	= portal vein
LA	= left atrium	UA	= umbilical artery
LV	= left ventricle	UV	= umbilical vein
PA	= pulmonary artery	IVC	= inferior vena cava
PV	= pulmonary vein	SVC	= superior vena cava
FO	= foramen ovale	A	= aorta
DA	= ductus arteriosus	DAor	= descending aorta
DV	= ductus venosus		

There are 2 umbilical arteries and 1 umbilical vein.

In man the situation is complex but similar basic mechanisms may provide the substrate upon which social and psychological conditioning lead to the development of adult psychosexual behaviour patterns.

The ability of the hypothalamus in adult animals to secrete a surge of gonadotrophin releasing hormone in response to oestrogen and progesterone priming is abolished by exposure to androgens during a critical period in fetal/neonatal development. Hence, ovulatory surges of luteinising hormone cannot be elicited in adult males or androgenised females. Experiments in animals suggest that regions of the limbic system concerned with the expression of psychosexual behaviour patterns also undergo an androgen-dependent differentiation.

FETAL CIRCULATION

The placenta acts as the organ of respiration for the fetus and blood with the highest O_2 concentration circulates to the brain (Fig. 12.3).

Blood is oxygenated in placenta and passes by umbilical vein (a) to liver (inner $\frac{2}{3}$) and (b) greater part circumvents liver by ductus venosus and enters inferior vena cava. Outer $\frac{1}{3}$ liver is supplied by portal vein. Fetal liver is partially functional.

The inferior vena cava receives blood from:

1. Ductus venosus (well oxygenated blood)
2. Hepatic veins (less oxygenated blood)
3. Hind body (very deoxygenated blood)

Oxygenated blood from inferior vena cava and deoxygenated blood from superior vena cava enter right atrium where:

1. Most blood (oxygenated) → FO → LA → LV → aorta to brain, coronaries, arms and head.
2. Reduced blood → RV → PA (90%) → DA to descending aorta. About 10% of blood from PA → lungs → PV → LA.

There is little intermingling of streams of oxygenated (posterior aspect) and reduced bloods in right atrium.

Flow through foramen ovale and ductus arteriosus short-circuits unexpanded lungs where vascular resistance is high.

The placental flow is about 55% of cardiac output.

The ventricles operate in parallel. The pressure in the PA exceeds that in the aorta; and the pressures in the right chambers are higher than those in the left.

FETAL RESPIRATION

The placenta is a feto-maternal organ with maternal sinuses through which maternal blood flows from uterine arteries to uterine veins.

Fetal capillaries in the chorionic villi dip into the maternal pools and gas exchange between the 2 discrete circulations takes place (Fig. 12.4).

The fetal capillary networks (cotyledons) are arranged approximately in parallel and are surrounded by blood in the maternal sinuses or intervillous spaces. Pulsatile jets of blood from the maternal spiral arteries perfuse the intervillous spaces, and a counter-current flow between the maternal and fetal circulations (not very efficient in man) is established. This creates a counter-current exchange system to facilitate gas, nutrient and waste-product exchange in the placental circulation.

The swirling of blood in intervillous spaces presumably causes PO_2 to be uneven in these sinusoids (cf O_2 diffusion in alveolar gas).

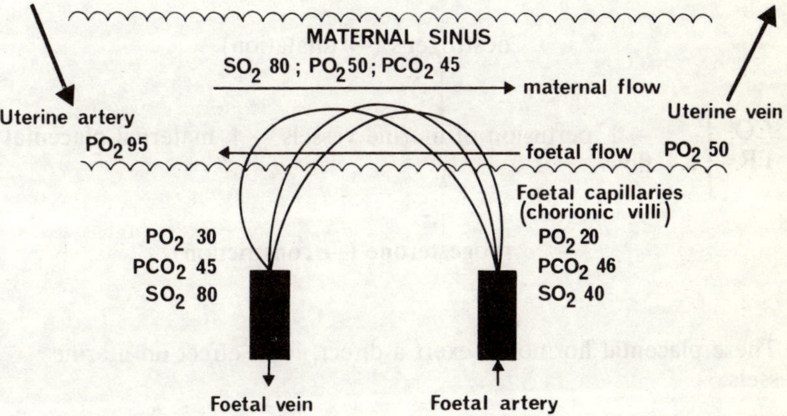

Fig. 12.4 Transplacental gas exchange near term

Key: The fetal and maternal circulations flow in opposite directions (an oversimplification of a very complex system)
SO_2 = % of saturation with O_2
Pressures are in mmHg

Transplacental gas exchange near term

The O_2 diffusion gradient is about 30 mmHg because diffusion is hindered by cells in chorionic villi (less permeable to O_2 than alveolar cells in the lung: blood-blood gas exchange barrier ~ 3.5 μm, cf ~ 0.5 μm blood-gas exchange barrier in lungs), and consequently O_2 equilibrium between the blood of fetal vein and maternal sinus is not attained (cf diffusion block in lung).

Nutrients and waste-products pass through fetal capillaries.

Maternal vein-fetal vein PO_2 difference is ~ 20 mmHg, and differences of this order are maintained even during administration of 100% O_2 (at 1 atmosphere pressure) to the mother (PO_2 in fetal vein rises only slightly); during uterine contractions in labour, and during altitude hypoxia (provided there are no maternal adverse effects).

Transplacental gas exchange is determined largely by the rate and distribution of the maternal and fetal blood flows so that gas exchange is chiefly flow limited rather than diffusion limited (as is gas exchange in postnatal lung).

CONTROL OF MATERNAL AND FETAL BLOOD FLOWS THROUGH THE PLACENTA

1. Maternal blood flow

The fetus is dependent on adequate uteroplacental perfusion.

oestrogens (→ dilatation)

↓ ⊕

↑ CO
↓ PR → ↑ perfusion of uterine vessels → ↑ maternal placental flow

↑ ⊖

progesterone (→ constriction)

These placental hormones exert a direct, local effect on uterine vessels.

2. Fetal blood flow

↑ CO
↑ BP (associated with ↑ fetal growth) → ↑ fetal placental flow

There are no known hormonal mechanisms for adjusting fetal blood flow through the placenta. However, the fetal sympathetic nervous system can cause a redistribution of fetal blood flow to the placenta (and cerebral and coronary vessels) when the fetal O_2 supply is compromised (e.g. on exposure to hypoxia or during extensive surgery).

CO = cardiac output
PR = peripheral resistance

MECHANISMS FOR INCREASING MATERNAL O_2 TRANSFER TO FETUS

As the fetus grows it progressively requires more O_2 and its O_2 needs are met by:

1. Increased uterine blood flow (20X);
2. Greatly increased fetal blood flow → ↑ removal of O_2 from each 100 ml maternal blood (since the ratio of maternal to fetal blood flow falls during pregnancy);
3. Fetal HHb mechanism (HbF has greater affinity for O_2 than HbA);
4. Increased fetal HHb concentration (40% higher than adult);
5. Double Bohr shift (giving up of CO_2 by fetal blood enhances O_2 unloading by maternal blood and O_2 loading by fetal blood).

At birth about —— 20% HbA / 80% HbF

HbF (α, γ polypeptide chains) takes up O_2 at low PO_2 due to poor binding of 2,3, DPG to α chain (HbA [α, β chains]).

EFFECTS OF LABOUR ON FETAL BLOOD PO_2. PCO_2, pH AND PLACENTAL PERFUSION

These parameters are relatively stable before the onset of labour and remain so during the 1st and 2nd stages of labour (a slight clinically unimportant fall in PO_2 may occur during 2nd stage).

The mean PO_2 in the uterine venous blood in pre-labour and during labour is steady, but it increases within minutes after birth (e.g. from 50 to 80 mmHg) as O_2 is not now delivered to the fetus.

Rhythmic uterine contractions decrease, and relaxations increase, maternal placental perfusion so that the mean perfusion is relatively stable.

During normal parturition, therefore, fetal asphyxia does not occur, but it is present in varying degrees immediately after birth until neonatal respiration commences and becomes functionally adequate.

If the maternal placental blood flow decreases then fetal hypoxia, hypercarbia and acidosis can occur.

PLACENTA

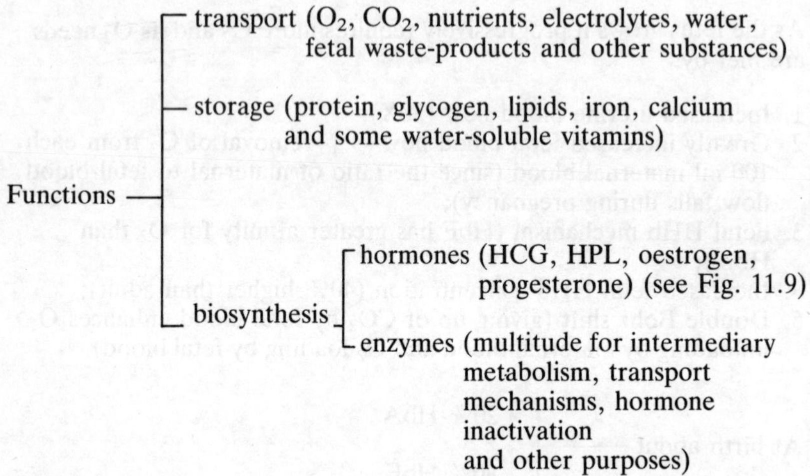

TRANSPORT FUNCTIONS OF PLACENTA

Transfer of substances across the placental membrane barrier is determined by the physicochemical and biophysical properties of the membrane and of the molecules to be transported, and by the volume of maternal and fetal blood flowing through the placenta: a decreased maternal blood flow results in a decreased transfer to the fetus.

Transport mechanisms in the placenta are complex and not well understood.

Carbohydrate transport

Glucose is the major metabolic substrate for the fetus. It can use, in small quantities, other energy substrates such as FFA, acetate, lactate, pyruvate and ketone bodies (e.g. when maternal ketosis exists).

In the fetus, the glucose level in the blood is lower than in the maternal blood; is largely controlled by the level of glucose in the maternal blood (i.e. homeostasis is largely regulated by the mother); and is only slightly influenced by fetal endogenous insulin.

Glucose crosses the placental membrane by facilitated diffusion, which is apparently independent of insulin (although placental insulin receptors of unknown physiological function have recently been described), and is decreased by maternal hypoxia (Fig. 12.5).

The placenta takes up glucose and converts it into glycogen and fat which are stored.

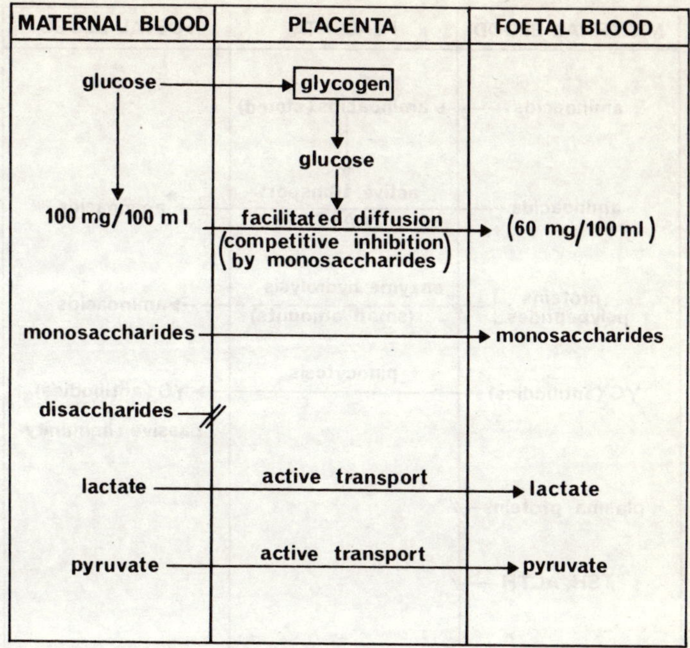

// = transport blocked

Foetal glucose derived from
- * maternal sources
- placental sources (glycogen)
- foetal sources (liver glycogen)

* = major source

Fig. 12.5 Transplacental transport of carbohydrate

Lipid transport

Fatty acids may be a more important source of energy for the fetus than was previously thought and, in the newborn, they compete with glucose as a major energy fuel.

During transplacental transport of fatty acids some are incorporated into neutral fat in the placenta and, in the fetus, some into neutral fat in the liver and adipose tissue. Fatty acids produced by lipolysis of neutral fat from these sources, and those which have crossed the placenta, can be oxidised to produce chemical energy.

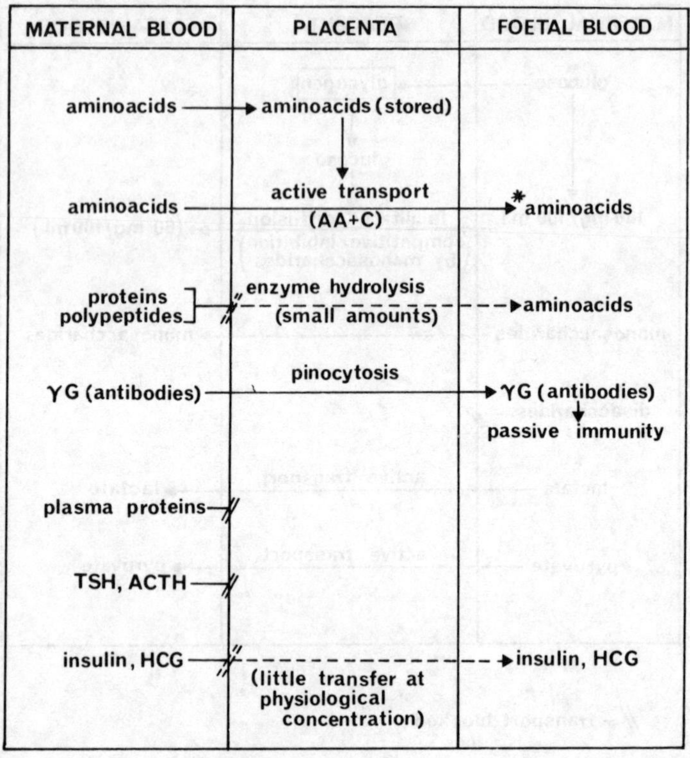

Fig. 12.6 Transplacental transport of protein

Key: AA = amino acids
C = carrier
* = in higher concentration in fetal than in maternal blood

The flux of fatty acids across the placenta is bidirectional and dependent on concentration gradients in the maternal and fetal blood (Fig. 12.7).

Towards the end of gestation, and during starvation in the mother, lipolysis in the maternal adipose tissue increases, and the supply of fatty acids to the fetus is enhanced.

The fetal and neonatal brain are able to use fatty acids and ketones as energy substrates. Transport of protein is shown in Figure 12.6, and transport of electrolytes, water and other substances in Figure 12.8.

AMNIOTIC FLUID

The amniotic fluid cushions the fetus against injury, provides a protective medium in which it can move, and fluid which it ingests.

The origin of amniotic fluid alters during fetal development: before mid-pregnancy it is largely a filtrate of fetal blood plasma;

after mid-pregnancy when the fetal skin becomes keratinised, it is mainly a product of fetal renal function (Fig. 12.9).

Constituents of amniotic fluid appear to be in continuous interchange with maternal and fetal circulations. Fetus swallows amniotic fluid which is absorbed into the blood and exchange of constituents takes place in placenta; this mechanism accounts for only some of the exchange which occurs, some is probably mediated via the amniotic membranes.

During fetal life the lungs are filled with amniotic fluid.

Uterine contractions during normal labour cause fluctuations in intra-amniotic fluid pressure which are equivalent to the fetus being submerged 2 to 3 feet under water with each contraction.

Intra-amniotic fluid infusion of amino acids for retarded fetal growth (~ 34 weeks) has recently been investigated.

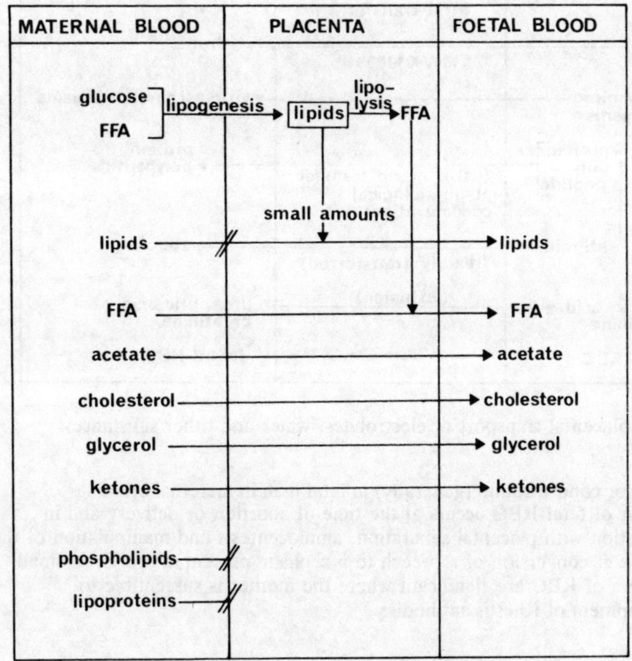

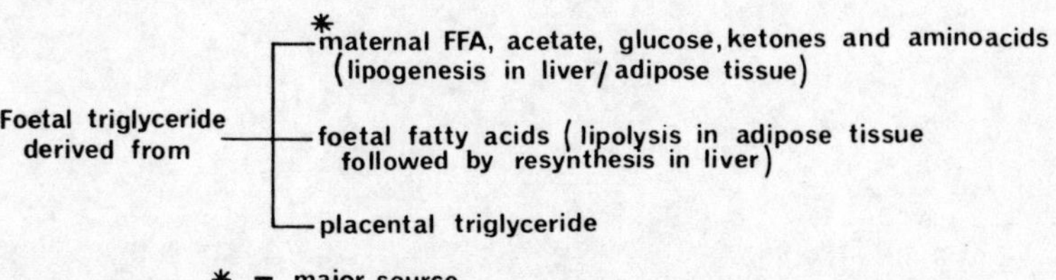

Fig. 12.7 Transplacental transport of lipid

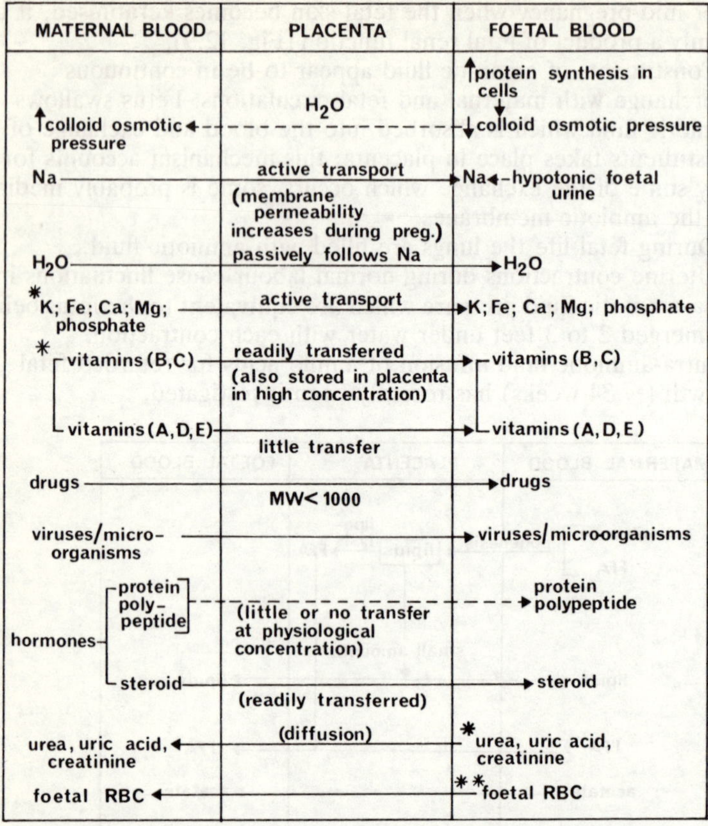

Fig. 12.8 Transplacental transport of electrolytes, water and other substances

Key: * in higher concentration (generally) in fetal than in maternal blood
** transfer of fetal RBC occurs at the time of abortion or delivery and in association with placental separation, amniocentesis and manipulation of the fetus (e.g. conversion of a breech to a cephalic presentation). Even small transfers of RBC are significant where the mother is susceptible to development of Rhesus antibodies

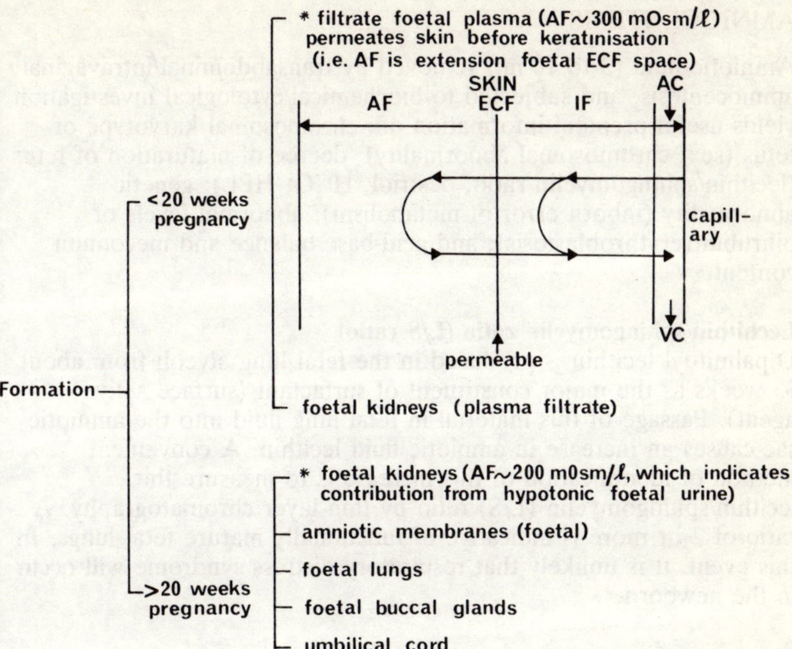

Fig. 12.9 Formation of amniotic fluid

Key: * = main source
AF = amniotic fluid
IF = interstitial fluid
AC = arteriolar capillary
VC = venous capillary

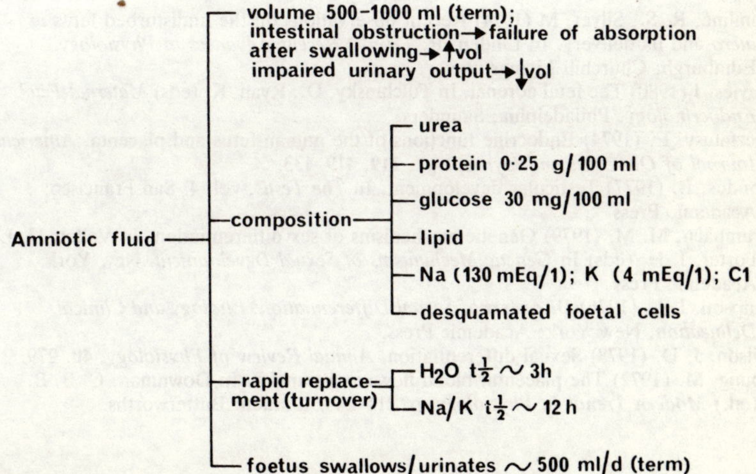

Fig. 12.10 Characteristics of amniotic fluid

AMNIOCENTESIS

Amniotic fluid (3 to 10 ml) removed by transabdominal/intravaginal amniocentesis, and subjected to biochemical/cytological investigation yields useful prenatal information on: chromosomal karyotype of fetus (sex, chromosomal abnormality); degree of maturation of fetus (lecithin/sphingomyelin ratio, oestriol, HCG, HPL); genetic abnormality (inborn error of metabolism); abnormal levels of bilirubin (erythroblastosis); and acid-base balance and meconium content.

Lechithin/sphingomyelin ratio (L/S ratio)

Dipalmitoyl lecithin is produced in the fetal lung alveoli from about 35 weeks as the major constituent of surfactant (surface active agent). Passage of this material in fetal lung fluid into the amniotic sac causes an increase in amniotic fluid lecithin. A convenient method of identification of this increase is to measure the lecithin/sphingomyelin (L/S) ratio by thin layer chromatography. A ratio of 2 or more is indicative of functionally mature fetal lungs. In this event, it is unlikely that respiratory distress syndrome will occur in the newborn.

FURTHER READING

Baum, M. J., Gallagher, C. A., Martin, J. T., Damassa, D. A. (1982) Effects of testosterone, dihydrotestosterone, or estradiol administered neonatally on sexual behaviour of female ferrets. *Endocrinology*, **111**, 773.

Belisle, S., Tulchinsky, D (1980) Amniotic fluid hormones. In Tulchinsky, D., Ryan, K. (eds) *Maternal-Fetal Endocrinology*. Philadelphia: Saunders.

Bühler, E. M. (1980) A synopsis of the human Y chromosome. *Human Genetics*, **55**, 145.

Comline, R. S., Silver, M (1974) Recent observations on the undisturbed fetus *in utero* and its delivery. In Linden, R. J. (ed.) *Recent Advances in Physiology*. Edinburgh: Churchill Livingstone.

Davies, J (1980) The fetal adrenal. In Tulchinsky, D., Ryan, K. (eds) *Maternal-Fetal Endocrinology*. Philadelphia: Saunders.

Diczfalusy, E. (1974) Endocrine functions of the human fetus and placenta. *American Journal of Obstetrics and Gynecology*, **119**, 419–433.

Gondos, B. (1977) Testicular development. In *The Testis*, vol. 4. San Francisco: Academic Press.

Grumbach, M. M. (1979) Genetic mechanisms of sex differentiation. In Vallet, H. L., Porter, I. H. (eds) In *Genetic Mechanisms of Sexual Development*. New York: Academic Press.

Simpson, J. L. (1976) *Disorders of Sexual Differentiation: Etiology and Clinical Delineation*. New York: Academic Press.

Wilson, J. D. (1978) Sexual differentiation. *Annual Review of Physiology*, **40**, 279.

Young, M. (1972) The placenta: blood flows and transfer. In Downman, C. B. B. (ed.) *Modern Trends in Physiology*, p. 214–244. London: Butterworths.

Multiple choice questions

1. In the fetus:

1. the most rapid weight gain is in last 8 weeks when fetal weight approximately doubles;
2. near term, the blood in the right ventricle has a higher PO_2 than that in the left atrium;
3. FFA are readily transported across the placental cells from maternal to fetal blood;
4. about the 36th week, the osmolarity of the blood equals that of the maternal blood;
5. iron is stored in the liver, and normally at term there is enough for haemoglobin synthesis for several months after birth;
6. concentration of haemoglobin in the fetus during the last trimester of pregnancy is usually higher than in the maternal blood;
7. meconium is produced entirely by the glands of the gut mucosa;
8. near term, bile pigments are not present in the meconium;
9. meconium may be passed into the amniotic fluid during episodes of fetal hypoxia.

2. In the fetus:

1. IgG antibodies are transferred to it across the placental cells probably by pinocytosis;
2. the heart begins to beat at 4 to 5 weeks;
3. near term, the per cent of O_2 saturation of blood in the fetal femoral artery approximates that in the cerebral arteries;
4. fat is synthesised from glucose, acetate and FFA which have been transferred from the maternal blood;
5. the uptake of iron, phosphorus and calcium is maximal in the last 6 weeks of full-term pregnancy;
6. cortisol secretion is dependent on ACTH since maternal pituitary hormones do not cross the placenta;
7. as the blood flows through the chorionic villi its pH becomes equal to that of the maternal blood;
8. vitamin D is stored in the liver, and near term there is normally sufficient to last the neonate for several weeks;
9. protein is synthesised from amino acids which have been actively transported across the placental barrier;
10. a reduced blood flow through the maternal placental sinuses during the last 6 weeks of full-term pregnancy will decrease fetal uptake of iron.

3. In the fetus:
1. fat-soluble vitamins are readily transported across the placenta;
2. glucose level in its blood will increase when the blood flow/min through the maternal placental sinuses increases;
3. insulin present in the blood is derived mainly from its own pancreas;
4. acid-base balance is regulated largely by the maternal lungs and kidneys;
5. plasma proteins are derived from maternal plasma proteins which have crossed the placental cells;
6. the heart rate at term is about 140/min and a decrease to less than 100/min suggests fetal distress;
7. phospholipids and lipoproteins freely cross the placental cells from maternal to fetal blood especially in the premature fetus;
8. most drugs cross the placental barrier to a significant extent;
9. blood glucose level will increase when the maternal blood glucose level increases.

4. In the fetus:
1. fetal haemoglobin (HbF) has a higher affinity for O_2 than HbA;
2. lipids freely cross the placental cells from maternal to fetal blood;
3. amino acids are in higher concentration in fetal than in maternal blood;
4. steroid hormones readily cross the placenta;
5. K^+ is in higher concentration in fetal plasma than in maternal plasma;
6. pressure in pulmonary artery is greater than that in aorta;
7. fetal haemoglobin has α and β polypeptide chains;
8. who is premature, severe hypoglycaemia (blood glucose $<$ 10 mg/100 ml) can occur;
9. the blood-brain barrier hinders transfer of ketone bodies to brain cells.

5. Glucose in the fetus:
1. is the major energy substrate;
2. is in higher concentration in fetal than in maternal blood;
3. diffuses (without energy expenditure) across the placenta;
4. is largely regulated by the glucose concentration in the maternal blood;
5. is metabolised, but this is not influenced significantly by fetal insulin;
6. can be derived from fetal hepatic glycogen;
7. can be derived from placental glycogen.

6. Differentiation of the reproductive tract in the fetus:
1. in the female results from the fetal ovaries producing substances which act on the primordial internal and external structures;
2. in the male is dependent on functioning fetal testes and end-organ sensitivity to substances produced by the testes;
3. in the XO karyotype individual is of the female pattern in both internal and external genitalia;
4. in the XX karyotype fetus exposed to excessive androgen during 7th to 10th weeks of fetal life is of the female pattern in the internal genitalia and of the male pattern in the external genitalia.

7. Normal intrauterine fetal growth:
1. depends upon an adequate supply of amino acids, fatty acids and carbohydrates;
2. in terms of fetal weight, is about 160 g/week in the last 5 weeks of pregnancy;
3. is associated with a placenta/fetus weight ratio of about 1 : 3 at 40 weeks of pregnancy;
4. can be augmented by intravenous infusions of lipids to the mother;
5. can be augmented by intra-amniotic infusions of amino acids.

8. With reference to chromosomal abnormalities in gynaecology:
1. amongst patients with primary amenorrhoea the likely incidence of chromosome abnormalities is 5 to 10%;
2. a patient who complains of primary amenorrhoea and who has a height of less than 145 cm is likely to have a chromosomal constitution of 45 XO;
3. a useful screening test for sex chromosome abnormalities is the counting of Barr bodies in cells scraped from the buccal mucosa;
4. a patient with an absent vagina, sparse axillary and pubic hair, but otherwise normal female body characteristics is likely to have a karyotype of 47 XXX;
5. patients with X-chromosome mosaicism are not capable of reproduction.

Answers

1. 1,3,5,6,9
2. 1,2,4,5,6,8,9,10
3. 2,3,4,6,8,9
4. 1,3,4,5,6,8
5. 1,3,4,5,6,7
6. 2,3,4
7. 1,2,5
8. 2,3

---

13. Perinatal and neonatal physiology

CARDIOVASCULAR SYSTEM

Before birth
- pr RA > pr LA
- pr RV > pr LV
- pr PA > pr aorta
  - high pulmonary R (deflated lung) (small cross-sectional area of lung circuit)
  - ↓ pulmonary BF
  - $\left[R \sim \frac{1}{XSA}\right]$
  - ↑ systemic BF
  - low systemic R (placenta) (large cross-sectional area of systemic circuit)
- work RV > work LV
- LVO > RVO (ventricles operate in parallel) (20%)

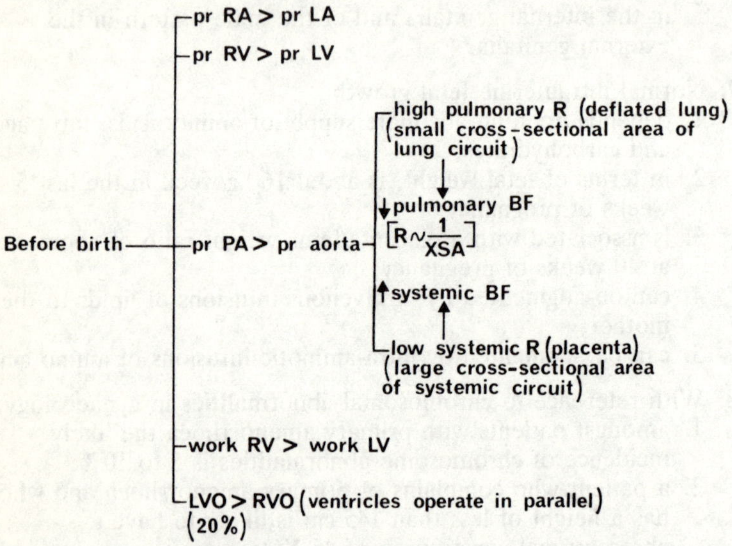

Fig. 13.1 Cardiovascular system before birth

Key: LVO = left ventricular output
R = resistance
RA = right atrium
RV = right ventricle
PA = pulmonary artery
BF = blood flow
XSA = cross-sectional area
~ = proportional to

The umbilical arteries cease pulsation several minutes after delivery and subsequently form the lateral umbilical ligaments. The umbilical vein remains patent longer than the umbilical arteries and, thus, permits placentofetal perfusion if cord is not tied immediately after delivery. Ultimately the closed vein fibroses to become the ligamentum teres. The fibrosed ductus venosus forms the ligamentum venosum.

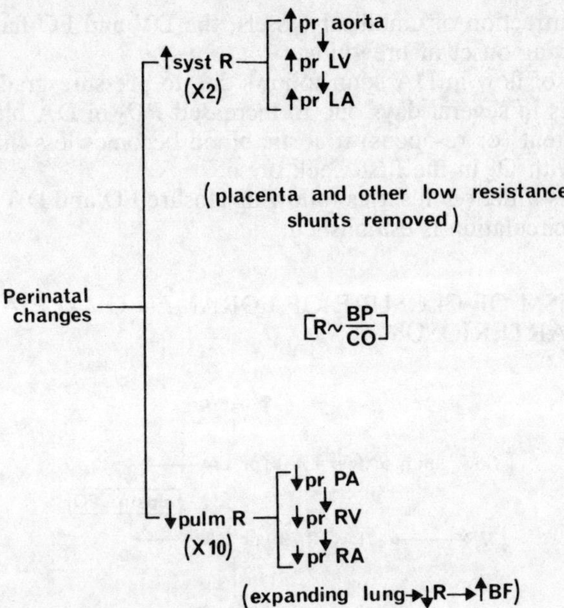

Fig. 13.2 Perinatal changes in vascular resistances

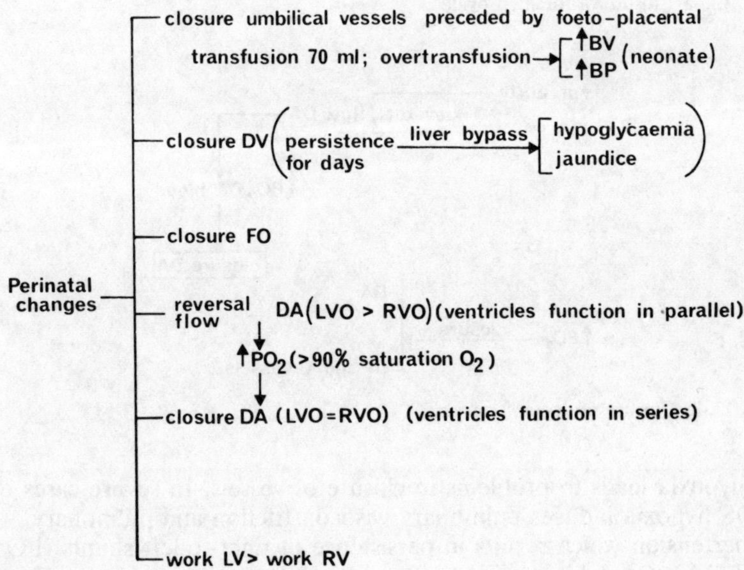

Fig. 13.3 Perinatal changes in vessels

Key: BV = blood volume

After contraction of umbilical vessels, the DV and FO flap-valve close following onset of breathing.

Reversal of flow in DA soon follows due to pressure gradient and vessel closes in several days due to increased PO_2 in DA blood. DA remains patent (or re-opens) if aortic blood becomes less than 90% saturated with O_2 in the first week or so.

Ventricles function in series following closure FO and DA and adult-type circulation is established.

MECHANISM OF CLOSURE OF FORAMEN OVALE AND DUCTUS ARTERIOSUS

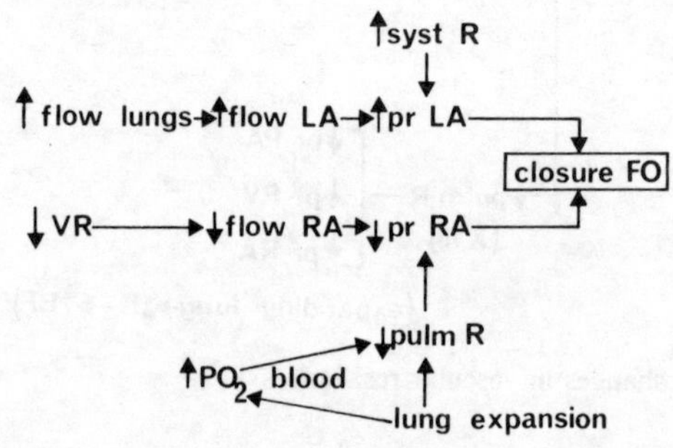

Fig. 13.4a Closure of foramen ovale

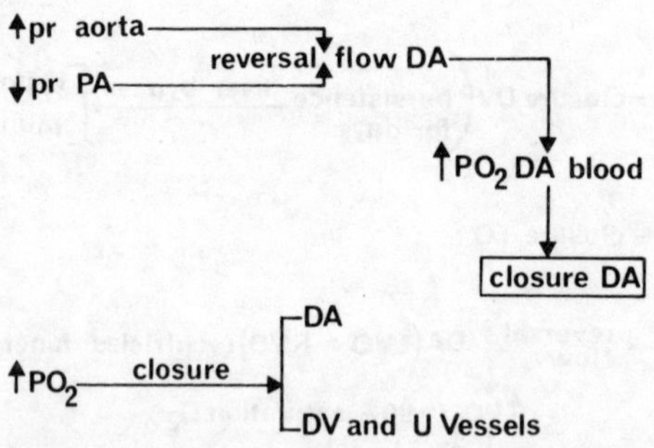

Fig. 13.4b Closure of ductus arteriosus

∴ hypoxia leads to problems in closure of vessels. In severe cases of RDS hypoxia induces pulmonary vasoconstriction and pulmonary hypertension which results in persistence of right-to-left shunts (FO and DA). Bradykinin and prostaglandins also assist in closure of ductus arteriosus.

RESPIRATORY SYSTEM

Before birth the fetus is exposed to a warm, aquatic, hypoxic and weightless environment and executes some rhythmic respiratory movements.

Fetal tracheo/bronchial/alveolar system contains fluid which is secreted or is an ultrafiltrate of plasma (or both).

Following birth the fluid in the lung is absorbed within several hours.

Bronchioles (and trachea) in newborn are about half diameter of adult, and the respiratory passages are lined with surfactant at alveolar air/fluid interface. Lung maturation (surfactant synthesis) is reflected in a L/S > 2 (see amniotic fluid)

Since: $P = \frac{2T}{r}$ (Laplace's Law)

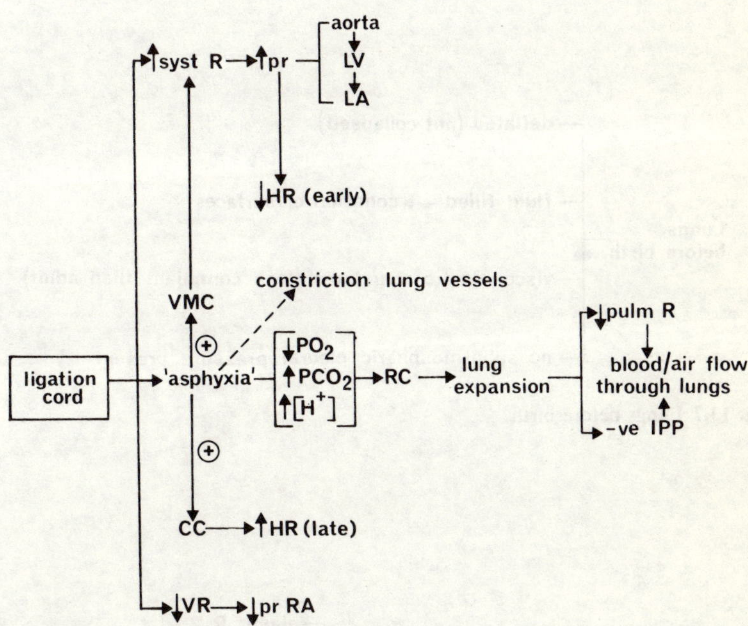

Fig. 13.5 Summary of cardiovascular changes following ligation of cord

Key: IPP = intrapleural pressure
VMC = vasomotor centre
CC = cardiac centre
RC = respiratory centre
HR = heart rate
VR = venous return

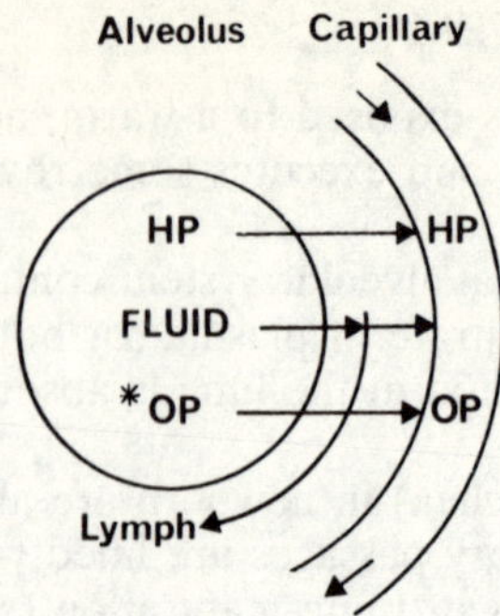

Fig. 13.6 Absorption of fluid from lungs

Key: HP = hydrostatic pressure
*OP = osmotic pressure (oncotic pressure); major factor

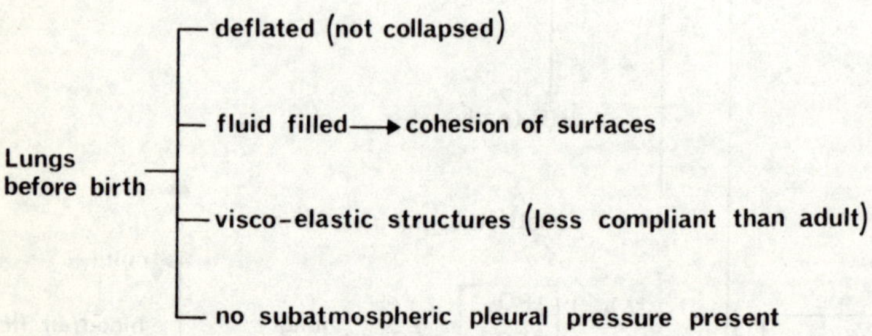

Fig. 13.7 Lungs before birth

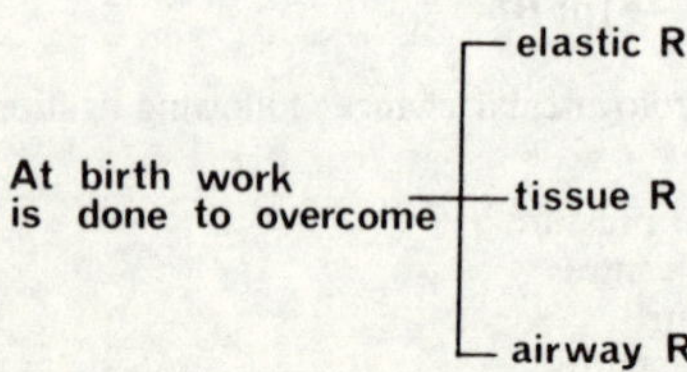

Fig. 13.8 Perinatal changes in lungs

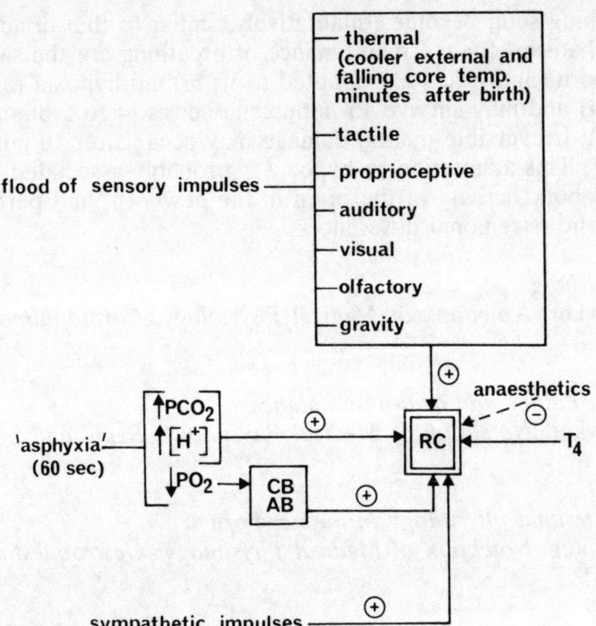

Fig. 13.9 Factors influencing onset of breathing

Key: CB = carotid body (chemoreceptors)
AB = aortic body (chemoreceptors)

It follows that, if surface tension is steady, the smaller the radius the greater the pressure required to open the alveoli.

Surfactant decreases the surface tension, and hence prevents collapse of alveoli during expiration.

Also, airway R $\alpha \frac{1}{D^4}$ (D = diameter airway)

Hence, the narrower conduits the greater the resistance to air flow.

Initially large inspiratory efforts (e.g. intrapleural pressure of minus 30 mmHg) are required to inflate the alveoli and within a few minutes, expansion of the lungs becomes progressively easier.

Thyroxine (T4) production increases in newborn and probably sensitises neurones of respiratory centre to incoming stimuli during the early days.

Establishment of functional residual capacity with onset of breathing keeps alveoli open, facilitates inspiration and expiration and militates against wide fluctuation in arterial PO_2 and PCO_2.

Breathing soon becomes qualitatively similar to that in adult, and the mechanisms for the maintenance of breathing are the same.

Newborn (and fetus) are adapted to withstand hypoxia (and asphyxia) and may survive 15 minutes apneusis (4 to 5 minutes fatal in adult); irreversible mental damage may occur after 10 minutes apneusis. This adaptation to hypoxia is probably associated with the low metabolic activity of the brain in the newborn, and perhaps this activity shows regional differences.

Neonatal lung
See Hawker, *Notebook of Medical Physiology: Cardiopulmonary*, Chapter 8.

Neonatal kidney and acid-base balance
See Hawker, *Notebook of Medical Physiology: Renal and body fluids*, Chapter 5.

Gastrointestinal physiology in the newborn
See Hawker, *Notebook of Medical Physiology: Gastrointestinal*, Chapter 9.

Total parenteral nutrition in the newborn
See Hawker, *Notebook of Medical Physiology: Gastrointestinal*, Chapter 9.

Neonatal monitoring
See Hawker, *Notebook of Medical Physiology: Cardiopulmonary*, Chapter 5.

FURTHER READING

Godman, M. J., Marquis, R. M. (1979) *Paediatric Cardiology: Heart Disease in the Neonate*. New York: Churchill Livingstone.

Grundmann, E., Kirsten, W. H. (eds) (1979) *Perinatal Pathology*. New York: Springer Verlag.

Heymann, M. A., Rudolph, A. M. (1975) Control of the ductus arteriosus. *Physiological Reviews*, **55**, 62–78.

Kumar, S., Rathi, M. (eds) (1978) *Perinatal Medicine: Clinical and Biochemical Aspects of the Evaluation, Diagnosis and Management of the Fetus and Newborn*. New York: Pergamon.

Murray, R. (ed.) (1979) *Respiratory Distress Syndrome*. Edison, N.J.: Medical Research Book Publishers.

Nathanielsz, P. W. (ed.) (1975) Perinatal research. *British Medical Bulletin*, **31**, 1.

Rudolph, A. M. (1979) Fetal and neonatal pulmonary circulation. *Annual Review of Physiology*, **41**, 383.

Multiple choice questions

1. Perinatal changes in the newborn occur as follows:

1. the systemic resistance increases and this increases the pressure in the aorta;
2. the pulmonary resistance decreases and this decreases the pressure in the pulmonary artery;
3. reversal of flow in the ductus arteriosus occurs and during this time the output of the left ventricle is equal to that of the right ventricle;
4. the changes in aortic and pulmonary artery pressures are reflected in pressure changes in the atria, leading to reversal of the pressure gradient across the foramen ovale resulting in its closure;
5. functional closure of the ductus arteriosus and ductus venosus may not occur if hypoxia persists for several days so that the aortic blood is only 80 to 90% saturated with O_2;
6. during the initial few breaths the intrapleural pressure falls by about 20 to 40 mmHg during inspiration;
7. the intrapleural pressure is below atmospheric at the end of expiration;
8. the body temperature falls soon after birth, and the neonate loses heat readily because of its large surface area/mass ratio;
9. after an initial fall in body temperature following birth the temperature begins to rise due to metabolism of brown fat.

2. In the newborn:

1. at term, vitamin K stores in the liver are sufficient for at least 4 weeks;
2. who is premature, the blood glucose level may fall to 20 mg/100 ml: because there is immaturity of gluconeogenic enzyme systems and increased sensitivity of cells to insulin;
3. compared with the adult, the kidney has a decreased capacity to concentrate the urine because there is inadequate development of the counter-current mechanism and decreased sensitivty of tubular cells to ADH;
4. fat is not metabolised for energy purposes before the first week;
5. who is premature, calcium absorption from small intestine is often deficient compared with older children and rickets can develop;
6. the thyroid does not concentrate iodide until after the first week;
7. there is adequate capacity to produce digestive enzymes;
8. compared with the adult, the stomach empties more rapidly and protein digestion in the small intestine may be delayed.

3. A week before birth (in the fetus):
1. the pressure in the right ventricle is greater than that in the right atrium;
2. the pressure in the ductus arteriosus always is greater than that in the right atrium;
3. the mean pressure in the right atrium is greater than the mean pressure in the left atrium;
4. the PO_2 in blood in umbilical vein is the same as the PO_2 in blood in maternal placental sinuses;
5. the oxygenated blood which enters the right atrium flows largely into the left atrium and not into the right ventricle;
6. about 90 per cent of the blood in the pulmonary artery short-circuits the lungs via the ductus arteriosus;
7. a glucose concentration of 60 mg/100 ml would cause irreparable brain damage;
8. liver and kidneys are functionally 'mature' organs;
9. the number of neurones laid down in the central nervous system is fixed and no additional neurones develop in the newborn;
10. irregular contractions of the uterus of about 30 mmHg intensity cause fetal hypoxia.

4. In the newborn:
1. compared with the adult, the kidney has a reduced capacity to regulate the osmolarity, volume and acid-base balance of the ECF, and this is exacerbated in the premature newborn;
2. when stressed by a water load a prompt diuresis does not occur and the urine is not appreciably diluted;
3. the cardiac output is about 1l/min at birth;
4. the kidney has a decreased ability to excrete H^+ ions and to acidify the urine for several days after birth;
5. the O_2 consumption at rest is about twice that of the adult on a w/w basis;
6. there may be a bleeding tendency in first few days because of inadequate production by liver of those clotting proteins which are dependent on vitamin K synthesis in colon;
7. synthesis of bile salts may be deficient, and hence fat absorption may be hindered; this is especially pronounced when the newborn is several weeks premature;
8. HCl is not produced by the parietal cells until after the first week.

5. In the perinatal period:
1. the plasma level of T3, as in the adult, results from that secreted by the thyroid and that produced in the tissues after monodeiodination of T4;
2. the placenta is impermeable to maternal TRF, TSH, T3 and

T4, but these compounds are present in the fetus from early gestation;
3. maternal hyperventilation will lower the maternal $PaCO_2$ but not the fetal $PaCO_2$;
4. with surfactant deficiency in the newborn the pulmonary vascular resistance is raised, pulmonary artery hypertension occurs and right-to-left shunts through the foramen ovale and ductus arteriosus can be produced;
5. hypoxia and acidosis in the newborn can lead to pulmonary vasoconstriction which produces pulmonary artery hypertension;
6. glucocorticoids (e.g. cortisol) given to the mother can induce surfactant secretion by the fetal lung;
7. when the maternal plasma FFA level increases more FFA crosses the placenta to the fetus;
8. FFA can pass in either direction across the cellular placental barrier.

6. In the perinatal period:
1. a factor in the production of physiological jaundice is a deficiency of ligandin;
2. ligandin complexes with bilirubin at the liver cell membrane and in this form bilirubin is transported to the microsomes;
3. transplacental O_2 exchange is chiefly flow limited rather than diffusion limited;
4. the PO_2, PCO_2 and pH of umbilical vein blood are relatively stable during normal first and second-stage labour;
5. the newborn responds to acute cold stress with an increase in plasma TSH, and in this respect differs from the response in the adult;
6. a postnatal rise in plasma TSH and T4 occurs, and this is so even when the ambient temperature is controlled;
7. the placenta is effectively impermeable to ADH, GH and ACTH, and their presence in fetal blood suggests their secretion by the fetus;
8. during induction of labour an overdosage of oxytocin to the mother can lead to maternal uterine spasm and fetal hypoxia and acidosis.

7. In the newborn:
1. one of the mechanisms of physiological jaundice is inadequate production of glucuronyl transferase for conjugation of bilirubin with glucuronic acid in liver;
2. the liver has a decreased capacity (compared to adult) for gluconeogenesis, ketone body formation, and deamination of amino acids;
3. humoral immunity is well established after the first week;

4. the glomerular filtration rate is about half that of the adult on a w/w basis;
5. who is premature the blood-brain barrier hinders glucose transfer to brain cells;
6. when stressed by a salt load there is a tendency to retain salt (and water) resulting in oedema;
7. the alveolar ventilation/min is about twice that of the adult on a w/w basis;
8. glucose in the extracellular fluid and that derived from liver glycogen is sufficient for its energy needs for only a few hours after birth;
9. ADH is not secreted until about the 4th week and hence the urine is hypotonic to plasma until the 4th week, especially in premature newborn;
10. a blood glucose level of 40 mg/100 ml does not impair cerebral function because brain cells can utilise energy substrates additional to glucose, e.g. lactate and ketone bodies.

8. With respect to perinatal physiology:
1. with maternal hyperthyroidism the infant may be hyperthyroid at birth;
2. with maternal post-thyroidectomy hypothyroidism the infant may be hypothyroid at birth;
3. with maternal hyperparathyroidism the newborn may have tetany;
4. with maternal diabetes mellitus the newborn may have impaired carbohydrate tolerance;
5. where glucocorticoids are administered in pharmacologic doses during pregnancy there is a risk of cleft palate in the infant;
6. iodine treatment of thyrotoxicosis in pregnancy may cause fetal goitre.

Answers

1. 1,2,4,5,6,8,9
2. 2,3,5,7,8
3. 2,3,5,6
4. 1,2,4,5,6,7
5. 1,2,4,5,6,7,8
6. 1,2,3,4,5,6,7,8
7. 1,2,4,6,7,8,10
8. 1,3,5,6

Symbols

$\uparrow$ = increases

$\downarrow$ = decreases

$\longleftrightarrow$ = unchanged

$\rightarrow$ = leads to; produces; causes

$\dashrightarrow$ = as above, but of secondary importance

$\xrightarrow{\oplus}$ = stimulates, increases

$\xrightarrow{\ominus}$ = inhibits; reduces

$>$ = greater than

$<$ = less than

$\sim$ = approximates

$\propto$ = is proportional to

[X] = concentration of X

Abbreviations

AA = Amino acid(s)
AB = Aortic body
ABP = Androgen-binding protein
AC = Adrenal cortex
Ach = Acetylcholine
ACTH = Adrenocorticotrophic hormone (corticotrophin)
ADH = Antidiuretic hormone (vasopressin)
AF = Amniotic fluid
AGS = Adrenogenital syndrome
AMP = Adenosinemonophosphate
ATP = Adenosinetriphosphate
A-V = Arteriovenous (difference or shunt)
BF = Blood flow
BMR = Basal metabolic rate
BP = Blood pressure
BSL = Blood sugar level (glucose concentration)
BV = Blood volume
cAMP = Cyclic adenosinemonophosphate
CB = Carotid body
CBG = Corticosteroid binding globulin (transcortin)
CC = Cardiac centre
CH_2O = Solute-free water clearance
CHO = Carbohydrate
CL = Corpus luteum
CLIP = Corticotrophin-like intermediate lobe peptide
CO = Cardiac output
CoA = Coenzyme A
COP = Colloid osmotic pressure
Cosm = Osmolar clearance
CRF = Corticotrophin-(ACTH-) releasing factor
CT = Calcitonin
d = Day
DA = Ductus arteriosus
DC = Drinking centre
DDAVP = 1-desamino-8-D-arginyl vasopressin
DHA = Dehydroepiandrosterone
DHT = Dihydrotestosterone
DIT = Diiodotyrosine
DNA = Deoxyribonucleic acid
DOPA = Dihydroxyphenylalanine
DP = Diastolic pressure
DV = Ductus venosus
ECF = Extracellular fluid
EOP = Effective osmotic pressure
F-1-6 diP = Fructose 1-6 diphosphate
$FADH_2$ = Dihydro flavine adenine dinucleotide

FFA = Free fatty acid(s)
FMP = Forward mobility protein
FO = Foramen ovale
F.P. = Flavoprotein
FRF = FSH-releasing factor
FSH = Follicle-stimulating hormone
GABA = Gamma-aminobutyric acid
G-6-P = Glucose-6-phosphate
GF = Graafian follicle
GFR = Glomerular filtration rate
GH = Growth hormone
GIF = GH-inhibitory factor (somatostatin)
GIP = Gastric inhibitory polypeptide
GIT = Gastrointestinal tract
GRF = GH-releasing factor
GRH = Gonadotrophin releasing hormone
GTP = Guanosine triphosphate
HCG = Human chorionic gonadotrophin
HDL = High density lipoproteins
HMS = Hexose monophosphate shunt
HP = Hydrostatic pressure
HPL = Human placental lactogen
HR = Heart rate
ICA = Islet cell autoantibodies
ICF = Intracellular fluid
IF = Interstitial fluid
IGF = Insulin-like growth factor
ILA = Insulin-like activity
IRI = Immunoreactive insulin
IVC = Inferior vena cava
JGA = Juxtaglomerular apparatus
kJ = Kilojoule
17KS = 17-ketosteroids (17-oxosteroids)
LA = Left atrium
LH = Luteinising hormone
LMP = Last menstrual period
LPH = Lipotrophic hormone
LRF = LH-releasing factor
LV = Left ventricle
LVO = Left ventricular output
ME = Median eminence
min = Minute
MIT = Monoiodotyrosine
mOsm/l = Milliosmoles/litre
MSH = Melanocyte-stimulating hormone
MW = Molecular weight
NA = Noradrenaline
NAD = Nicotinamide adenine dinucleotide (coenzyme 1) (DPN)
NADP = Nicotinamide adenine dinucleotide phosphate (coenzyme 11) (TPN)
$NADH_2$ = Dihydro (reduced) NAD
$NADPH_2$ = Dihydro (reduced) NADP
NL = Neural lobe
NS = Neural stalk
NSILA = Non-suppressible insulin-like activity
O = Oestrogen
OP = Osmotic pressure

P = Progesterone
PA = Pulmonary artery
PBI = Protein bound iodine
PEP = Phosphoenolpyruvate
PFK = Phosphofructokinase
PG = Prostaglandin
PIF = PRL-inhibitory factor
PK = Pyruvate kinase
PP = Pulse pressure
PR = Peripheral resistance
PRF = PRL-releasing factor
PRL = Prolactin (lactogenic or luteotrophic hormone)
PTH = Parathyroid hormone
PV = Pulmonary vein
PVN = Paraventricular nucleus
PZ-CCK = Pancreozymin-cholecystokinin
R = Resistance
RA = Right atrium
RBC = Red blood cell
RC = Respiratory centre
RMP = Resting membrane potential
RNA = Ribonucleic acid
RV = Right ventricle
RVO = Right ventricular output
SO_2 = O_2 saturation %
SON = Supraoptic nucleus
SP = Systolic pressure
ST = Seminiferous tubules
SV = Stroke volume
$t_{\frac{1}{2}}$ = Biological half-life
T3 = Triiodothyronine
T4 = Thyroxine
TBG = Thyroxine binding globulin
TBPA = Thyroxine binding pre-albumin
TBP-T4 = Thyroxine bound to thyroxine binding proteins
TCA = Tricarboxylic acid cycle
TCH_2O = Solute-free water reabsorption
TeBG = Testosterone-oestradiol binding globulin or SHBG (steroid hormone-binding globulin)
Tm = Tubular transport maximum
TRF = TSH-releasing factor
T/S = Thyroid to serum concentration ratio
TSAb = Thyroid-stimulating antibodies
TSH = Thyroid-stimulating hormone
UDP = Uridine diphospho (as in UDP-glucose)
$\dot{V}$ = Volume/min
VIP = Vasoactive intestinal peptide
VR = Venous return
XSA = Cross-sectional area

Index

Note: page numbers in italics refer to a diagram

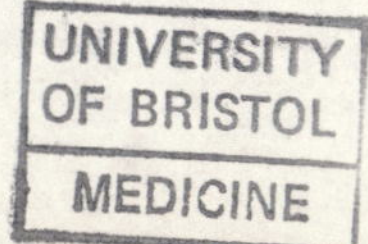